46. 20

P9-DNS-401

Understanding Babylonian Astronomy

Hedonizing Technologies

Paths to Pleasure in Hobbies and Leisure

RACHEL P. MAINES

The Johns Hopkins University Press
Baltimore

© 2009 The Johns Hopkins University Press
All rights reserved. Published 2009
Printed in the United States of America on acid-free paper
2 4 6 8 9 7 5 3 1

The Johns Hopkins University Press
2715 North Charles Street
Baltimore, Maryland 21218-4363
www.press.jhu.edu

Library of Congress Cataloging-in-Publication Data

Maines, Rachel, 1950–
Hedonizing technologies : paths to pleasure in hobbies and leisure /
Rachel P. Maines.
p. cm.
Includes bibliographical references and index.
ISBN-13: 978-0-8018-9146-5 (hardcover : alk. paper)
ISBN-10: 0-8018-9146-9 (hardcover : alk. paper)
1. Hobbies—History. 2. Amusements—History.
3. Technology—Economic aspects—History—
Popular works. I. Title.
GV1201.5.M35 2009
790.1'309—dc22   2008043993

A catalog record for this book is available from the British Library.

*Special discounts are available for bulk purchases of this book. For more information,
please contact Special Sales at 410-516-6936 or specialsales@press.jhu.edu.*

The Johns Hopkins University Press uses environmentally friendly book
materials, including recycled text paper that is composed of at least 30 percent
post-consumer waste, whenever possible. All of our book papers are acid-free,
and our jackets and covers are printed on paper with recycled content.

For Daryl M. Hafter,
with as many thanks as there are stitches
in the Bayeux Tapestry

CONTENTS

This book project grew out of a lifelong interest in needlework, home preserving, and cooking. My great aunt, Elsie Niles, and her friend and neighbor Ella Reynolds, on the other side of Aldrich Street in Hope Valley, Rhode Island, in the early 1950s, were responsible for introducing me to the pleasures of handcrafts as leisure activities. My parents, Natalie and Donald Petesch, have always encouraged my reading, writing, and scholarship, and inadvertently inspired my interest in food and cooking by their joint distaste for working in the kitchen. Rebellion against one's parents in such matters can be very satisfying, even to the parents in question.

My late friend Catherine Mary Gatto de Oliver (1950–2007) was an inspiration as an indefatigable hedonist and hedonizer of everyday life. She would have loved this book. I owe much to Judith Ruszkowski and Karenna LaMonica for their friendship and support over the past thirty years.

The greatest debt I incurred in the research and writing of this work, however, is to the Cornell University Department of Science and Technology Studies (STS), which has provided me with a professional home since 2004. My association with STS has afforded access to Cornell's vast library resources in print and online, including those of the Rare and Manuscript Collection, which are beyond praise in their scope, quality, condition, quantity, and availability to researchers.

For intellectual support and exchange of ideas, I am particularly grateful to my STS colleagues Trevor Pinch, Michael Lynch, Ronald Kline, and Peter Dear, as well as graduate students Honghong Tinn, Katie Proctor, Janet Vertesi, Joseph "Jofish" Kaye, Benjamin Wang, Victor Marquez, Darla Thompson, Kathryn Vignone, Robert Schombs, Hansen Hsu, and Hannah Rogers. When I presented this project in the Social Sciences Research Group series in STS in the spring of 2008, comments by Katie Proctor on gardening, by Mike Lynch on fishing and its gear, and by Bruce Lewenstein, Suman Seth, and Steve Hilgartner on definitions of hedonization all helped sharpen the focus of this work.

Jonathan Coopersmith, of Texas A&M University, all-round fine scholar and great guy, organized a lively session at the Society for the History of Technology (SHOT) meeting in Amsterdam in 2004, where I presented some very early and less-than-coherent results of this research project. Merritt Roe Smith and graduate student Kieran Downes, of the Massachusetts Institute of Technology, and David Lucsko, of the University of Detroit–Mercy, included me in their session on hobbies, including audiophilia and ham radio, at the 2007 SHOT meeting, attracting a crowd that spilled out into the hallway. Many thanks to them and to the session participants, especially Joseph Corn, Beverly Sauer and Kristen Haring. Thanks also to my SHOT colleague and friend Deborah Warner, of the Smithsonian Institution, for great feedback and for laughing in all the right places. Daryl Hafter, now emeritus from Eastern Michigan University, to whom this book is dedicated, introduced me to SHOT in 1976, setting me on the path to a doctoral degree and a life of scholarship.

Rebecca Slayton and Robert McGinn invited me to speak in the Stanford University Science, Technology, and Society colloquium series in February 2008, a thoroughly enjoyable experience that was a great help in refining some of the ideas in this book. Charles Perrow and Robert Proctor made especially helpful comments, and Joseph Corn elaborated on the illuminating analysis he had presented at the SHOT session in 2007.

I am indebted to my former editor (now publisher) at MarketResearch.com, Donald Montuori, for the opportunity to study the role of prepared food in modern American lifestyles, and to Edward McCambridge, of the Chicago firm Segal, McCambridge, Singer & Mahoney, for his insights into the complex relationship between fishing and the tying of flies.

As always, my first reader, harshest critic, husband, and main squeeze, Garrel Pottinger, contributed his insights, insisted that I clarify concepts like the distinction between plain and fancy needlework, and provided many nights of excellent cooking and cozy snuggles during the four years I worked on this book.

I owe much as well to several anonymous referees for the Johns Hopkins University Press, who helped me identify the proper shape for my exceptionally unwieldy subject.

And finally, let's hear it for library catalogers everywhere. We couldn't do any of this without them.

Hedonizing Technologies

Introduction

In 1978 family sociologists Jay Mancini and Dennis Orthner published a paper in the *Journal of Sex Research* reporting the results of a survey of leisure preferences among couples in a southwestern city. The top pick among the men was predictable: 45 percent preferred to spend their spare time having sex, with "attending athletic events" and "reading books" coming in second and third. The women, however, preferred books to sex: 37 percent would rather be reading. Sex ran a poor second at 26 percent, nearly tying with activities that did not even register on the male leisure scale: needlework and sewing at 25 percent.[1]

Because the authors do not tell us the survey's margin of error, it is a moot question whether sex really won the race for second place on the distaff side of the family. The study enjoyed its fifteen minutes of fame in the form of a brief but spirited debate in newspaper and other periodical editorials, and in indignant letters to advice columnists by men who took women's unflattering preferences as a personal criticism, as indeed they may well have been.[2]

A Gallup poll in March 1977 had also ranked needlework high among women's leisure choices. In answer to the question "What is your favorite way of spending an evening?" a predictable 30 percent of both sexes chose watching television, with reading a faraway second at 12 percent for men and 17 percent for women. Needlework was the favorite activity of 8 percent of the women in this sample and, not very surprisingly, of less than 1 percent of the men, so small a number that Gallup did not even bother to report it. Sex was not offered as an option on the survey checklist, and the response "home with family" is probably not an adequate surrogate.[3]

The Gallup study showed that women did needlework across all income categories, and that they were only slightly more likely to do so if they were rural homemakers rather than urban professionals. Needlework, once a necessity for the poor and a luxury for the rich, has become in the last two centuries a rewarding hobby for large numbers of women—and sometimes men—of all social classes: the discount store clerk who crochets afghans, the schoolteacher who knits, the stockbroker who works cross-stitch samplers.

None of these artisans has any need to produce clothing or other goods, or even to spend much time repairing them. Their engagement with needlework is not a practical matter. It is now a leisure activity like sex, gourmet cooking, and amateur photography, and like them its technology has adjusted to a market that privileges the enjoyment of the process over the goal of efficient production. The Industrial Revolution has made leisure needlework a technology of pleasure, along with many other preindustrial activities once performed by necessity and now marketed to consumers as leisure pastimes, including weaving, woodworking, ceramics, painting, hunting, fishing with hook and line, antique auto restoration, the making of soap and candles, cake decorating, food preserving, and horticulture. The success of "home improvement" and "do-it-yourself" enterprises such as Home Depot, and Martha Stewart's print and broadcast obsessions with domestic crafts, are only the most recent manifestations of our historical pleasure in technologies we own and control with eye and hand. In the past ten years, there has been an explosion of published and reprinted titles on pickling, preserving, drying, and other methods of "putting food by," despite the ubiquity of grocery stores. Like the taste of grain, the sensation of manipulating raw material satisfies us at a deep (and possibly genetic) level, resonating with adaptive behaviors dating at least to the Paleolithic. Apparently, many of us like to do things with our hands, whether we need to do them or not.[4]

Although technological change since the eighteenth century has produced conditions in which hobby and leisure activities are economically accessible to nearly everyone in Western industrial democracies, the phenomenon of what we may call "leisure work" is much older. Gardening, hunting, and fishing became hedonized by the wealthy in antiquity as soon as it was unnecessary for these individuals to hunt or gather their own food. In Britain and Europe, leisure hunting by the nobility was by the Middle Ages at the very latest an activity virtually irrelevant to nutrition; at the same period, needlework among noblewomen was already hedonized, not only at home but also in convents, where entire communities set aside time to enjoy the quiet and satisfying rhythm of passing a needle through cloth, making the elaborate, richly decorated garments and church ornaments that the nuns themselves did not use.[5] Their modern sisters would crochet doilies and potholders by the dozen in much the same way, and in pursuit, I shall argue, of the same kind of pleasure.[6]

CHAPTER ONE

What Is a Hedonizing Technology?

Any technology that privileges the pleasures of production over the value and/or significance of the product can be a hedonizing technology. One would intuitively suppose that some technologies would resist hedonizing—coal mining and air traffic control, for example, and ironing and darning among domestic activities—but one would be wrong. All of these work algorithms have their counterparts among hedonized activities. There are hobby coal miners and air traffic control simulators, and while contributors to the website www.extremeironing.com are not really much interested in pressing wrinkles out of cloth, nonetheless, some people darn and others iron for pleasure.

Novelist Diane Schoemperlen vividly evoked the nuances of this type of recreational work in her 2002 *Our Lady of the Lost and Found*, describing the "relaxing and comforting" movements of her iron, the smell of the fabric, and the "serious and stalwart heft" of her obsolescent iron. We see here the appreciation of good tools that are enjoyable to handle, the joys of the process, and the irrelevance of the product that characterize technologies of pleasure.[1] Later, Schoemperlen's narrator confesses that she sometimes takes "half a dozen perfectly pressed shirts from my closet" and irons them again just for the druglike sedative effect.[2] The experiences she describes resonate with what Samuel Florman has immortalized as those of engineers *en rapport* with their tools and machines.[3] Schoemperlen's quietly passionate relationship with her iron would no doubt be familiar to ham radio and hobby woodworking enthusiasts, although most of these are male. Gender socialization clearly influences what hobbies an individual chooses, but the technological satisfactions are definitely cognate.

Needlework is a paradigmatic example of this process of hedonization, with the kinds of nuances from one technique to another that are characteristic of technologies making the transition from drudgery to delight as they lose their practical relevance to production. This book is about how technological change

Hobby coal miners. Participants in this leisure activity must receive the same safety training as professional coal miners. Photo by Chris Murley, reprinted by permission.

in the larger culture has hedonized hobbies and leisure activities at the individual level, using five of the needle arts as central examples and case studies: home sewing, quiltmaking, knitting, embroidery, and the sister arts of crochet, tatting, and lacemaking. We shall examine how the removal of all need to make, sew, or repair textiles in the home freed artisans—most, but not all of them, women—to enjoy and explore the aesthetic, sensual, intellectual, and emotional rewards of needlework, and even to find in it a form of salutary escapism from the troubles of ordinary life, whether pre-, proto- or postindustrial. We shall observe how these arts have differed from each other in the time, the place, and the manner of their hedonization in Western history. We shall also explore parallels in other hedonized crafts and leisure activities, and survey briefly other craft and leisure activities that have parallels with the hedonization of needlework, cooking, and gardening, including those associated mainly with men and boys.

The comedian George Carlin has observed that "people who are not free are the most free about what they have left to be free about." He was speaking of African American leisure-time music and humor, but the same principle can be applied to other groups and activities. While the choice of leisure activity may be

Extreme ironing is clearly more about the "extreme" than about the ironing. Here, ironist Troy Wallett wins the Rowenta Trophy in 2003, pressing a shirt while suspended over the Wolfberg Cracks in Cederberg, South Africa. Photo by Gordon Forbes, reprinted by permission.

limited by economic or gender-role constraints, hobbyists of both sexes and all races seem to cherish the freedom to choose some, possibly very small, corner of their lives over which they have full and unchallenged creative control. Genealogy, for example, one of America's most popular hobbies, is a kind of hedonized historiography, free of the disciplinary, peer review, and publication standards to which professional historians are required to submit.

There is an atavistic or archaizing element in many leisure technologies that seems to enhance the pleasure of participation. Both camping and outdoor cooking, for example, are enjoyable in part because they are inconvenient and labor-intensive, inviting the full involvement of one's attention and thus the exclusion of ordinary daily concerns.[4] Although some affluent consumers install what are in effect alfresco professional kitchens on their decks and patios, it is demonstrable from economic data on the sales of grills, barbecue gear, and charcoal that Americans (and Australians) of all socioeconomic strata are seriously addicted to cooking the occasional meal in the tradition of Neolithic hunter-gatherers.[5]

Recreational barbecue technology can be either archaizing or state-of-the-art: re-enactors and members of the Society for Creative Anachronism favor wood cookfires lit with iron and flint; Southern Californians and members of the deck-and-patio demographic favor gas grills with electronic controls and refrigerated compartments for ingredients. Many hobbyists are attracted to "professional quality" tools and equipment: restaurant-grade cookware, for example, and full-scale lathes and machine tools for home workshops.

Historians generally, and historians of technology in particular, have had very little to say about the development of technologies in which the primary product is individual and/or communal pleasure. Sexuality, cooking, ham radio, photography, and hobby drag racing are the most obvious counterexamples, although it could be argued that the last is closely connected to public leisure. Of these, only sexual technologies for private pleasure have received significant attention from historians. The public consumption of leisure and recreational activities—travel, amusement parks, the theater, brothels, taverns, motion pictures and so on—have all been explored by historians in greater or lesser depth, perhaps at least in part because these involve identifiable and documentable economic enterprises.[6] It is much harder to document and quantify assertions like that of Ruth Schwarz Cowan that historically, people simply enjoy their food more when they eat it in the privacy of their own homes. Anecdotally and experientially, it is obvious that they do; Cowan uses traditional resistance to cooperative kitchen and mass feeding schemes, which are clearly more efficient from the point of view of production, to demonstrate how jealously these private pleasures are guarded.[7] At this writing (fall 2006), cookbook sales are booming, as are kitchen renovations, but only 53 percent of all meals eaten in the American home are prepared there. Of these "home-prepared meals," a significant percentage is represented by sandwiches and bowls of cold cereal.[8] Almost half of the average family's food budget is now spent on the purchase of prepared food, either from the grocery or restaurant carryout, or on site in restaurants. Clearly, most of those who still cook from scratch are doing so because they like to do it, and are investing in the tools, materials, and workspaces associated with culinary pursuits as a hobby.[9]

Historian of technology Reese Jenkins points out in his 1975 *Images and Enterprise* that the private pleasure associated with family snapshots—meaningless to anyone who does not know the subjects and lacking any kind of aesthetic or economic value—helped make photography a major industry in the twentieth century.[10] Amateur photographers were and are economically important, but we rarely see historical studies of how successfully different photographic technologies in the roll-film era enhanced the private pleasure of this large group of consumers.[11] Historians of photography tend to see developments from the point of view of professionals, who are usually seeking efficient, cost-effective techniques, marketable reproduction quality, and image permanence.[12] Hobbyists—amateurs—are looking for something else.

This "something else" is missing also from studies of technologies in which private pleasure is explored on the margins rather than in the mainstream. Robert Post's *High Performance* is an exceptional book about automotive history in its inclusion of hobby dragsters and other car enthusiasts.[13] Amateurs, except for those involved in the very early stages of technological development, are rarely taken seriously in studies of, for example, railroad and aviation history, even when it would be appropriate to do so. In part, this is because amateurs and hobbyists complicate the analysis: these consumers have criteria for technological success that are quite different from those that historians are accustomed to considering. For example, Walter Vincenti's study of fixed and retractable aircraft landing gear takes the position that the configuration of gear should contribute to speed and fuel efficiency. He considers only the use of aircraft to transport goods, people, and weapons from one location to another, omitting to mention the counterexample of aerobatic flying, a technology of pleasure, in which the lighter weight of fixed gear make them the standard configuration.[14] Even the term *technology* is, in the minds of many scholars, associated with productive work and economic value.

While there are few generalizations that apply to all hedonizing technologies, there are commonalities among many of them. Needlework, for example, shares with lithography, ceramics, soapmaking, fishing, hunting, and many other hobbies and arts the characteristic of having become more attractive as a leisure activity as its tools and methods become less practical as means of profitable production. Knitting socks by hand, now a popular hobby, was once a professional occupation for both sexes and a home duty for many wives and mothers. After the introduction of knitting machinery, and the entry of increased numbers of women into the labor force outside the home, it gradually ceased to be an economically efficient means of producing footwear. While the transition of knitting from necessity to hobby was already well under way in the nineteenth century, it was not complete until after the First World War.[15]

Similarly, in the late nineteenth century, after rail transportation had driven most commercial stagecoaching from the roads of Britain and the United States, coaching became a popular, if expensive, hobby for "fast" young urban men, to the great amusement of William Thackeray.[16] In the twentieth and twenty-first centuries the manual automobile transmission, happily discarded by most of us as soon as automatics became affordable, is the pride and hallmark of a person, often but not always male, who enjoys the robustly interactive element of driving a car.[17] Traditional color lithography as a graphic medium was taken over by artists and hobby artisans after it became obsolete in commercial printing.[18] Home candlemaking is currently on the rise as a hobby, even though most of us have not needed the products to light our homes for more than a century and a half.[19] Even shaving has become hedonized, with badger shaving brushes, straight razors and their strops, and shaving soap in cakes making a market comeback at luxury prices.[20]

Once production shifts to industrial methods, the leisure consumer is free to seek pleasure in the older handcraft technology. Typically, the technology itself enters one or more paths to pleasure as the market recognizes hobby demand: tools and materials are designed for comfort, beauty, and satisfaction.[21] Both needlework tools and those of hobby woodworking have undergone this transition, to name only two of many possible examples.[22] Fountain pens, considered obsolete as a production technology for writing, are selling at four-figure prices, through catalogs like Levenger's, to people who simply enjoy the process of forming words with ink on paper and are willing to pay a premium for the pleasure.[23] In the 1950s, the late Shelby Foote reportedly wrote his three-volume 1.5–million-word history of the Civil War with a dip pen, eschewing the then-dominant writing technologies—the manual typewriter and the fountain pen—thereby lending a new meaning to the term "belletristic history."[24]

Hedonizing technologies do not follow the same paths as those for market production because they do not need to satisfy the same criteria. Steel knitting needles are more durable, faster in working, and smoother than wooden needles, but they do not feel as good in the hands. Retractable landing gear make aircraft more efficient and faster in flight, but hobby aerobats eschew them because they make the plane heavier and less able to "dance" in the aerobatic box. Many hobby technologies, including woodworking, needlework, and home preserving, are consciously archaizing, moving "backwards" on the path of technological "progress." In the case of the hobbies, the diversity of paths that flourish as their tools, materials, and designs democratize is one of their most persistent characteristics.

What makes a technology hedonize? Why do we cling to "obsolete" but enjoyable technologies for our leisure moments? What elements of craft and leisure technologies make them enjoyable? Are any technologies un-hedonizable? Is industrialization of production a necessary condition for hedonizing the crafts? Is hedonization a cross-cultural phenomenon, or is it an artifact of the industrializing North Atlantic, as one referee suggested? The hedonizing potential of craft technologies—and their market implications—raise questions that historians and economists alike have left almost entirely unaddressed.

I have chosen the neologisms *hedonize* and *hedonizing* to denote the process by which the leisure practitioners of technologies change their focus from production and efficiency to personal pleasure, as archaizers shift their focus to the styles and techniques of the past. *Hedonistic* was proposed as an alternative, but hedonism is a quality of human (and sometimes animal) character, not of technologies. I also wanted to capture the intentionality of the change; according to the *Oxford English Dictionary (OED)*, the ending *-ize*, which has a quite respectable Greek pedigree, may be used either transitively or intransitively, and has a history of use for neologisms of the kind I propose. The *OED* adds that *-izer* is an "agent noun," and it is this agency to which I intend to call attention.[25]

Theories of Leisure

Mark Twain's Tom Sawyer tells us that work is whatever "a body is obliged to do" and play is "whatever a body is not obliged to do." The Industrial Revolution, by making nearly all consumer goods cheaper and more readily available, and by expanding leisure, opened a much larger space in Western life for hobbies, the work-like play that "a body is not obliged to do." Among more formal analysts of leisure, the psychologist Abraham Maslow asserts that the pleasures of creativity are luxuries to which we turn when our "deficit needs"—physical security, food, shelter, and so on—have been satisfactorily met.[26] Leisure crafts and hobbies like knitting, ham radio, model-making, and aerobatics are social and economic luxuries, and they typically arise from the divergence of production needs from craft traditions. We seem to begin to enjoy tasks, and to insist on re-skilling them, when it is no longer economically necessary to perform them ourselves. But not all tasks, even within a particular craft area such as needlework and sewing, undergo this hedonization at the same rate.

Conversely, as almost everyone who has ever turned her leisure craft into a business can testify, formerly enjoyed tasks can undergo *agonization*, to coin a necessary antonym, when one is compelled to perform them day after day in order to get one's livelihood.[27] When one gets one's bread from craft, at least part of the motivation is extrinsic; for true hobbies, the leisure theorists tell us, the motivations and satisfactions must be at least primarily intrinsic; they must be, as anthropologist Clifford Geertz tells us, "deep play."[28]

Theories of leisure have struggled to accommodate hobbies like needlework, both because they are productive—it is, after all, needle *work*—and because early theorists in this field, particularly Thorstein Veblen, were notoriously ill-attuned to the nuances of home as a leisure workplace.[29] As Leslie Bella has pointed out, leisure theorists before the 1970s avoided discussions of home crafts as hobbies by arguing either as Veblen and Lorine Pruette did, that to the extent that women did not leave home to work, all their time counted as leisure, or, at the opposite extreme, as Sebastian De Grazia said of ancient Greek women, that because these individuals had no freedom from required tasks, they could not have experienced leisure.[30] Even the economists, whose efforts recall Lewis Carroll's "seven maids with seven mops sweeping for half a year," have come to grief in the project of assimilating hobby artisanship into a model of home production, similar to doing laundry and cooking family meals.[31]

By the time Joffre Dumazedier wrote in 1967 that women's "semileisure" at home might occasion some "semipleasure" in their activities there, just as men's home maintenance and gardening hobbies did, it was apparently beginning to dawn on at least some social scientists that what counts as a leisure activity cannot be defined by reference to the activity, but only by the freedom of the individual to choose it, and by his or her enjoyment of it for its own sake.[32] Attempts to document historical hobby activities using works of art encounter just this problem: it is impossible to tell by looking at a picture whether or not the artisan is having fun.

Needlework is an especially tricky activity to categorize, because to nonparticipants, it is by no means obvious what distinguishes work that is socially or economically required, and that which is chosen by the artisan. Documentary evidence, such as diaries, letters, and even convent account books, illuminate this distinction, however: needleworkers and indeed all hobbyists know when they are working and when they are having fun.

Fiction by and about women is a useful source of data on how the idea of leisure needlework changed between the eighteenth and the twenty-first centuries; novelists from Fanny Burney to James Joyce and Margaret Atwood have depicted social and economic distinctions among the needle arts.[33] In the fiction of Anthony Trollope (1815–82) and Margaret Oliphant (1828–97), for example, women made and mended clothing because they were required or expected to; hobby needlework—usually embroidery or lacemaking—was typically a source of pride and pleasure.[34] The marketing of needlework materials, from medieval peddlers to A. C. Moore, has also reflected the distinction between necessity and leisure, with seductive use of color and texture to appeal even to those with little to spend on luxury materials. The appeal to pleasure can be documented in needlework advertising to all classes in Europe and America since the eighteenth century. Whether the buyer is ordering hand-painted knitting wool from a website or buying acrylic yarn in Wal-Mart, advertising and merchandising of product are addressed to the sensorium of the artisan.[35]

The subject of hedonizing technologies has been rendered exceptionally murky by attempts to reconcile hobbies and leisure with Marxian and other theories of capitalism, with results that can only be described as comical. Steven M. Gelber, for example, after asserting in his 1999 *Hobbies: Leisure and the Culture of Work in America* that hobbies did not exist before the nineteenth century, goes on to argue that "leisure is part of the cultural scaffolding that has been constructed to hold up Western free-market capitalism."[36] Theories of this kind rest on the premise that hobbies and leisure activities constitute either a rebellion from, or an accommodation to, the alienation of work thought to be associated with industrialization, or perhaps both at different levels of context.[37] The obvious objection is that both predate by many centuries either capitalism or industrialization, and it is not immediately obvious how, for example, Seneca's gardening, the needlework of Mary Queen of Scots, or Izaak Walton's fishing can be construed as having either supported or rebelled against "Western free-market capitalism."[38]

As we have seen, Marxian and related theories seem to accommodate predominantly male forms of leisure more readily than those pursued mainly by women, perhaps because, historically, men have oftener had workplaces outside the home, making it easier to determine when they were working and when they were "at leisure." For example, Hugh Cunningham's 1980 *Leisure in the Industrial Revolution*, a study of British working class leisure between 1780 and 1880, tells us much about male social activities such as ratting, drinking, music halls, gambling, cockfighting, and cricket; a little about gardening; and nothing whatever about hobby needlework, although its popularity among British working-class women in the nineteenth century is amply documented.[39] Even Jennifer Hargreaves' 1989 article on British women's leisure tells us much of sport and nothing of hobby crafts, although her work is useful for its careful attention to the questions of "social and economic constraints . . . (what is *possible*), the connection with ideology and symbolic forms of control (what is *permissible*), and self-determination and the articulation and fulfillment of desire (what is *pleasurable*)."[40]

Sociologists Harold Wilensky and Stanley Parker are responsible for the concept of leisure as either a compensation for what is lacking in working life or a spillover of its culture, ethics, and values.[41] Home hobbyists or leisure artisans, in this model, are looking both for some reward denied to them in the modern workplace, which sociologists seem to agree is a place of alienation and acutely felt powerlessness, and for the socially reinforcing "valorization" of productive leisure: idle hands, we are told, are the Devil's workshop. Social scientists seem also to agree that reduced working hours outside the home and economies of scale in manufacturing and distribution have made hobby tools and materials and the time to enjoy them accessible to a much larger percentage of the population, a conclusion in which I concur. They seem, however, more or less united in a belief that there is something exceptionally alienating about work in the industrial and postindustrial world that has driven the growth of hobbies in the twentieth and twenty-first centuries. John Clarke and Charles Crichter call this trend the "Protestant leisure ethic."[42]

All of these theories of historical exceptionalism, from Veblen's through Wilensky's to Gelber's, fail to explain why the two most popular hobbies in modern America—gardening and textile crafts—were also the two most popular hobbies in Homer's Greece, Seneca's Rome, and Norman England.[43] The influence of industrialization has been to democratize preferences that appear to have been enduring and even, in some cases, cross-cultural. Embroidery, by definition a decorative rather than a structural technique, appears as hedonized in Nazca Peru as it was in the nineteenth-century British aristocratic household, although of course we do not have access to the Nazca makers' views as we do with the British examples.[44] Even gender lines have been routinely crossed: sailors at nearly all periods of recorded history, who have had little space, money, or opportunity for other hobbies, have taken up fiber arts, including embroidery, in their long shipboard leisure hours, as well as artisan carving of wood, ivory, and bone.[45] Surviving artifacts and documents also attest to hobby crafts in detention camps and prisons.[46]

Technological change and industrialization have also wrought changes that make the home a leisure environment: lighting and heat in every room, basements and garages for workshops and tool storage, reclining chairs within easy reach of our workbaskets, computers on which to exchange patterns and ideas for hobby projects, to name only a few.[47] Some of these innovations in living space have themselves been driven by the demand for leisure-work environments, such as that oddly named entity, the one-and-a-half car garage, built to accommodate a home workshop along one wall.

Part of the difficulty of reconciling serious hobbies like needlework and gardening with theoretical distinctions between work and leisure is that they are worklike in their requirements of time, labor, and skill. Johan Huizinga, author of *Homo Ludens,* notes this quality when he asserts that "strenuosity in the work of plastic art obstructs the play-factor," though he acknowledges that some artisans must enjoy their work.[48] Hobbies like gardening and needlework can be forms of nadification—nest-building—which for humans has always existed in a gray area between labor and rest. Political scientist Sebastian de Grazia, writing of work and leisure in 1964, notes his own confusion about do-it-yourself work for both sexes. After a brief mention of home sewing and decorating, he goes on to lament "the rash of do-it-yourself activities—the plumbing, wiring, carpentry, painting, landscaping (to put it euphemistically)—that the man of the house suffers heroically."[49]

When we dig in our own soil to plant flowering trees, build a backyard barbecue, or crochet antimacassars for our furniture, are we enjoying the activity for the sake of the simple physical and creative pleasure we feel in it? Improving our property? Making our personal environments more comfortable and aesthetically pleasing? Building up our self-esteem and enhancing our reputations for competence? Killing time? Finding ways to bond with others interested in either similar goals or similar activities, as in the cases of quilting and model railroading? All of the above? I do not propose to disambiguate these issues, but to suggest that the pleasure we feel in hedonizing technologies assimilates elements from all of these motivations and others as well. Part of our pleasure is derived from the recognition of productive virtue, even as we wallow in the voluptuous sensations of handspun wool yarn sliding over wooden needles, the brilliant hues and rich flavor of homemade jelly, or the lush scents of our own humus turned with a spading fork.

The tools that aid and abet these virtuous pleasures are judged by artisans who do not answer to supervisors, have no profit motive, and who vote for or against new tools, materials, and techniques with their buying power as consumers. To this extent, at least, the leisure theorists are correct in perceiving a connection between capitalism and hobbies. Their mistake lies in failing to recognize that it has always been buyers who have ruled this technological market—at Ruth Cowan's "consumption junction"—and who have shaped its destiny from Homer to Home Depot.[50] When we can afford the leisure—the eight-hour work days with their weekends off that industrialization and the labor movement have bought for us, the latter with committed activism over more than a century—we satisfy our craving for Geertz's "deep play" with tools and materials of our own choosing, and support the businesses that produce, sell, and market these commodities.

Unlike industrial workers, hobby artisans have complete control over what tools and materials they use, and since efficiency of production and marketability of product are rarely issues, they buy or make the tools and materials they most enjoy using. Leisure theory and sociological sources illuminate the kinds of pleasures artisans experience in their craft, which include control and mastery, the beauty or the individuality of the product, the sensual enjoyment of the task—what leisure theorist Mihaly Csikszentmihalyi calls "the flow," the escapist quality of immersion in voluntarily chosen work, and socialization opportunities with other crafts enthusiasts.[51]

Gender and class socialization obviously has a strong influence on choice of craft; for instance, needlework is typically a female activity, while woodworking is more often male. As the historian J. P. Toner has observed of ancient Rome, "Clearly, leisure has to be felt, but as sensual experience is intimately connected with the conventional social manners of its expression. This leads to an acceptance that the pursuit of enjoyment—in short, doing what one wants, likes and finds pleasurable—is not a purely personal concern."[52] If one is to be at one's ease in a leisure activity—particularly a hobby which, like needlework, gardening, or woodworking, requires for full enjoyment an absorption in the task that excludes ordinary concerns—it cannot be an activity that routinely brings serious sanctions from the community. A Saudi woman, for example, might make a hobby of driving a car, but she cannot be at ease in it in her own country in the same way that she could fearlessly devote herself to an afternoon of embroidery.

Even when not actually life-threatening, gender pressure on leisure choices may require some personal insouciance—or other formidable qualities—to resist, as when former football star Roosevelt ("Rosey") Grier took up canvas embroidery in the 1970s. Only an exceptionally brave or foolhardy gender-role conservative would have challenged the 300–pound, six foot five Grier on his choice of hobbies to his face.[53] Most real and fictional invaders of gendered hobby preserves, however, have had to rely less on implied physical threat than on simple persistence or extenuating circumstances to make their choice a success: a female model railroader who inherited her father's or husband's layout, for example, or a male invalid whose knitting or embroidery was tolerated because of his infirmity.[54] While this inflexibility about gendered leisure crafts is changing, no observant visitor to a yarn store, or to a model train shop, would argue that it is changing very visibly or very fast.

Hedonizing Technologies in the Leisure Universe

Most of the technologies of leisure craft activities originated as means of performing productive work. One might argue, for example, that because tatting has never been a measurable economic activity and has been a hobby for nearly all of its two-hundred-year existence, a tatting shuttle is not a means of performing work. In fact the instrument's evolution from the making of knots in more economically significant contexts, such as fishing net and ship's rigging, can be readily traced.[55] The dominant trend of these technologies is that they have been co-opted by leisure artisans, who have exerted consumer pressure on their form in such a way as to enhance the pleasure of using them.[56]

This pressure can, as we have seen, take completely different forms for the same general type of activity without conflict, just as the market for crunchy peanut butter leaves plenty of room for consumers of the smooth product. In production settings, a single technological choice can become the dominant means of performing work: retractable landing gear in aircraft, for example, and ring spinning in textile production. Hobby technology, on the other hand, adapts to the preferences of individual consumers much more readily and shows much more diverse characteristics in use. It can adapt to so-called "markets of one" in which hobby tinkerers play a significant role.[57] As a rule, it is less capital-intensive than its industrial counterparts, although in woodworking and automobile restoration expenditures per capita in production and hobby shops can be comparable. The same can be said of home sewing on a sewing machine (although not of the procedures for cutting fabric) and of hobby photography and home movie-making.

While I have been addressing mainly cases of hedonizing technologies whose tools and/or techniques are obsolete for industrial production, industrial obsolescence is not always the case. Two or more entirely different methods of practice for hobby technology may co-exist for what can be described as components of the same task. For one textile artisan, the pleasure of craft may consist in spinning her own yarn on a drop spindle, and for another, programming the embroidery settings for her high-tech sewing machine. Cooking outdoors can be, and is, hedonized at both ends of the technological spectrum, from roasting meat on sticks over a wood fire to state-of-the-art gas grills with every conceivable modern convenience.

In most of these characteristics, hedonizing technologies differ from other kinds of recreational technology, such as amusement devices, toys, and sports equipment. Amusement devices, whether MP3 players, computer games, or Ferris wheels, are produced by (and, in the case of Ferris wheels, operated by) persons employed to perform this work. Typically, either the device itself is purchased and used with minor modifications by the consumer, or the experience produced by the device is "consumed" by a purchaser who has little to do with its operation. With the notable exception of components purchased by hobby stereo tinkerers, machines like record players, televisions, bongs, radios, MP3 players,

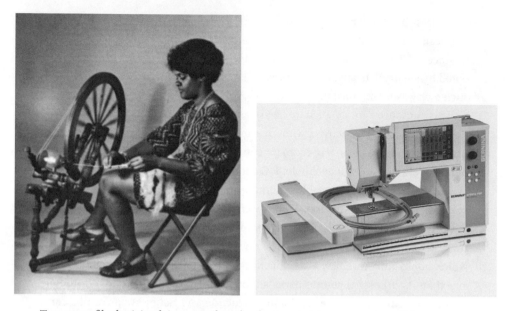

Two ways of hedonizing leisure textile technology: archaizing versus state-of-the-art. *Left*, a handspinner operates a foot-powered spinning wheel, c. 1970; *right*, Bernina's most sophisticated home sewing and embroidery machine, the Artista 730, 2008. Allen Fanning, *Handspinning Art and Technique* (New York: Van Nostrand Reinhold, 1970); photo by Bernina International AG, reprinted by permission.

vibrators, and the like are of the first type. Motion pictures shown in movie theaters, carnival rides, tour buses, and similar types of recreational technology fall into the second category.

Toys for children represent a fairly well-understood type of recreational technology, usually some sort of miniaturized or less-dangerous version or representation of the artifacts and circumstances of adult life. We have innumerable historical examples of toy soldiers, animals, baby humans (dolls), weapons of all kinds, transportation and construction equipment, costume, cooking gear, and the like.[58] These objects are more like practice tools for the realities children will encounter as adults than technologies that have evolved for grown-up artisan pleasure. Yet some toys, such as looper looms and spool knitting kits, mirror adult hobbies or, as historian of technology Svante Lindqvist has pointed out about Meccano (what Americans call Erector sets), prefigure and intersect with adult artisanship and technical interests in complicated ways.[59] Yuzo Takahashi has made the same observation regarding radio hobbyists in postwar Japan.[60]

Sports equipment, too, presents categorization challenges in this context. Clearly, a golf club or a catcher's mitt is a tool, but except for those owned and

Two ways of hedonizing outdoor cooking: archaizing versus state-of-the-art. *Left,* Mrs. Cecil Hull of Welch, West Virginia, cooks fish fillets the old-fashioned way on 4 July 1955 at Cranberry Campground, Monogahela National Forest, West Virginia; *right,* Frontgate Professional Grill, 2008. Photo by Leland J. Prater © Corbis, reprinted by permission; photo by Frontgate, reprinted by permission.

used by sports professionals, are they not also toys for adults? I shall argue later on that hunting and fishing for sport, as well as camping and paintball, are hedonized technologies, but are these fundamentally different from athletic activities as recreation?

All of the boundaries that I have suggested here are evidently porous: what of bicycles, for example, which have been hobby technologies and articles of sports equipment, as well as means of transportation, since their invention? Do not many bicyclists spend hours tinkering with their equipment? Is a child's bicycle a toy? Is a college student's disposable camera a recreational technology of the MP3 player type, or a toy that happens to produce a product? Surely it is only a hobby technology if the student is serious about photography; is such an individual likely to buy a disposable camera?

Even toys and accessories for pets and other animals merit consideration in the schema of recreational and hedonizing technologies. The issue is fairly straightforward with, for example, amateur equestrians: the tack, the gear, and even the clothing are hedonized from earlier functional traditions of horse transportation.[61] The issue becomes more vexed, however, with the tools, toys, and gear associated with smaller companion animals, such as pet collars, tug toys, food dishes, and litter boxes, technologies into which innovations are introduced every year. Are not these artifacts intended to enhance our enjoyment of companion animals? There is no doubt that a horse collar is technology; the issue is cloudier, however, with collars for cats and dogs. A fish tank in the home is clearly a technological artifact—indeed, many are complex biochemical and mechanical systems requiring significant commitments to daily care. Do we want to say that these devices are hedonized versions of scientific aquaria?

However inclusive or exclusive a definition for hedonizing technologies might ultimately be, I intend to limit my discussion here to a fairly narrow range of hobby and leisure activities, starting with textile crafts practiced primarily but not exclusively by women, and comparing their paths from production to leisure to those of hobbies traditionally associated with men, such as fishing and hunting for sport. Both of these sports become pastimes contemporaneously with leisure embroidery in Western life, as expensive and socially significant activities for an aristocracy with more leisure than it knew how to use.

CHAPTER TWO

Leisure and Necessity

C. M. Woolgar's 1999 study *The Great Household in Late Medieval England* portrays a social stratum that was, in the period 1066–1500, so thoroughly separated from productive work that mealtimes for members of the nobility could run to three hours. Servants, mostly male even in the Great Wardrobe, labored night and day to produce food, clothing, ale, and beer, clean laundry, and every other necessity of life for a handful of women and men, some of the more elevated of whom were so closely attended that they did not even need to wipe their own derrières. When not eating or drinking or courting each other's wives or fighting their enemies or each other, men of this class hunted and hawked, read books, gambled, played or listened to music, or, more rarely, went fishing for sport.[1]

Medieval Noblewomen as Hobby Artisans

Most historians of leisure and hobbies seem to regard the embroidery of medieval noblewomen and nuns as if it were work, possibly because many of them do not know how to make the distinction between plain sewing and fancywork, which will be explained later in this section, or perhaps because it is not perceived as parallel to the largely male activities of hunting, hawking, and fishing. What seems to have largely escaped scholarly attention about this use of time by rich women of the Middle Ages is that it was both unnecessary on any practical grounds, and socially approved.[2] Moreover, it permitted display of wealth not only in the richness of tools and materials employed in it, but also in the employment of fashionable artists as designers of patterns. On these grounds, I argue that medieval noblewomen embroiderers represent the first hobby needleworkers in Western history.

The hypothesis of Steven Gelber and others regarding the non-existence of hobbies before the nineteenth century has at least one element of intuitive appeal: the rise of hobby-related publications in that century made leisure activities much easier to distinguish from their murky backgrounds of work and necessity. Since the principal defining elements of a leisure activity are that one does not have to do it and that one enjoys the process, our lack of information regarding what preindustrial individuals considered necessary and whether or not they enjoyed what they were doing is a serious handicap to hobby and leisure historiography.

Leisure gardening and hobby farming by those who could afford professional gardeners seem definitely to have existed in antiquity, but we know little about whether its practitioners—particularly Romans like Seneca who were very conscious of their appearance of conformity (or lack of it) to traditional virtues—actually enjoyed the activity for its own sake, or whether it was a kind of social chore, comparable to attending a socially necessary party at which having a good time is not the point. Hunting for sport is known to have been a standard activity of affluent men in Greek and Roman times, but again, we cannot determine the social pressures to participate or observe the states of mind of the participants.[3] The elder Pliny, arguably antiquity's most prolific font of unreliable information, tells us that gardening was a source of pleasure for the likes of Epicurus and Cato, but we have corroborating first-person accounts of gardening from neither.[4] The same difficulties apply to the embroidery of the female aristocracy: we know that they spun, wove, and embroidered, and that in strictly economic terms they did not need to do so, but we know nothing whatever about their perceptions of social necessity, nor how they felt about conforming to it. It is written documentation that makes the mind of the individual in history accessible to us, and often even that is ambiguous, as in the case of Seneca. Hunting, like gardening, can be a genuinely pleasurable form of leisure or, for men of elite classes in various historical times and places, a social obligation.[5] I shall return to the hedonization of hunting in the context of modern leisure in Chapters 4 and 5.

Although the questions of social pressure and personal enjoyment persist, individual agency regarding hobbies and leisure began to emerge from the murk of necessity in the Middle Ages, and the technologies of these hobbies began to take on some of what we shall come to recognize as the characteristic paths to hedonization.

Historically, most women have been "obliged," in Tom Sawyer's sense, to do a great deal of needlework, particularly plain sewing, darning, and mending, which in the industrial era is now economically and socially unnecessary. Anthropologist Elizabeth Barber argues persuasively that textile making became associated with women at least in part because it is compatible with the historically omnipresent responsibility for child-herding and with the care of the sick.[6] Diaries, letters, and fictional accounts by women show clear evidence, however, that many women and some men (including a number of sailors and sailmakers) chose certain needlework tasks for pleasure even when other tasks, like mending, were performed as work.[7]

The distinction between plain and fancy needlework has for the most part dropped out of common twenty-first-century parlance, principally because the parallel processes of industrialization and hedonization have made all needlework optional. The element of necessity formerly associated with plain work has been entirely eliminated. As historical categories, however, plain and fancywork are surprisingly difficult to define; as in many other craft contexts, it is easier to point to examples than to formulate comprehensive definitions. The needlework of necessity—constructing and mending clothing and household textiles, knitting and darning of hosiery and other garments, and the laundry marking of linens with embroidery to identify the owner—were all considered plain work.[8]

While hemming of household linens and clothing was plain work, fancy hemstitching with drawn-thread embroidery was fancywork. Knitting was plain work unless it involved lace, the making of silk purses, or the embroidery method called "duplicate darning" over the knitted stitches. Although linen marking employed counted-thread embroidery stitches, it was considered plain work because it was done for functional rather than decorative purposes. Outside the home, fancywork was any kind of needlework that required skills beyond the basic stitches of plain sewing: running, back, the flat-felling techniques used for hemming, overcasting, and whipstitch for rolled edges. Buttonhole making, for example, paid more than hemming because it required more skill; professional embroiderers usually earned more than dressmakers.[9]

Embroidery is probably the most common and earliest type of hedonized needlework technology, emerging by the late Middle Ages as a recreational activity for women who had the resources and the leisure. Even within this medium and among the aristocracy, some embroidery tasks were undertaken as work and others for pleasure. There is no reason to believe, for example, that the Herculean task of textile press agentry represented by the Bayeux Tapestry was regarded by its professional embroiderers as anything but an interminable, if lucrative, chore.[10] What few accounts we have by women themselves, such as those of Christine de Pisan (1364–1431) and Elizabeth de Burgh (mid-fourteenth century), suggest that when noblewomen had leisure time that was under their own control—as opposed to, say, the obligation to plan and preside over those three-hour dinners—they read or were read to, listened to or made music, and embroidered, often in groups.[11]

Women and girls of the merchant and artisan classes were also taught sewing and embroidery in medieval times, but the focus of this training was not for leisure but as a kind of human capital formation: the trainee would have a potential source of income either for her husband if he needed or wanted it, or for herself if widowed.[12] Beguines—lay women who chose to live a religious life in the kind of single-sex communities of which at least one is still flourishing at the Beguinage in Amsterdam—seem also to have treated needlework as a marketable skill rather than as a leisure activity, causing no end of trouble for the male-dominated embroiderers' guilds, whose prices they of course undercut.[13]

Noblewomen, a tiny minority in households dominated by their male counterparts and by male servants, could decently involve themselves in recreation by riding, walking in the garden, falconry, reading, music, visiting, or embroidery.[14] Reading, music and visiting were all compatible with needlework, as in a great household plenty of noble and educated persons-in-waiting were available to read aloud or play or sing to one or more women working at their embroidery frames. While the embroidered clothing and household textiles of aristocratic English households in the late Middle Ages are known to have been made almost exclusively by professional embroiderers (many of them male), women of leisure nevertheless embroidered in the absence of any pressing necessity to do so. This absence of need is well-documented: for example, when King Canute's wife Aelgifu gave embroidered altar cloths to abbeys, and Edward the Confessor's wife Edith embroidered robes for her husband, in both cases contemporaneous records noted that the work was done "with her own hands," to distinguish it from the commissioned needlework of the kind donated to Winchester by Queen Aelfflaed in the early tenth century.[15]

When women of the aristocracy entered religious life, it was not unusual for them to choose either an order in which needlework was an established area of specialization (as distilling was for some monasteries), or to choose the embellishment of church textiles over possibly less enjoyable aspects of convent work, such as tending the sick, gardening, cleaning floors and privies, cooking, or doing laundry.[16] St. Etheldreda, an English abbess of the mid-seventh century, is said to have entered religious life mainly so that she could spend her days working elaborate embroideries in gold and silk, the most luxurious and costly of needlework media.[17] Even in convents, embroidery was an elite occupation; medieval historian Shulamith Shahar says of this that "in the richer nunneries all household chores were done by lay sisters or servant women and the nuns occu-

Embroidery was hedonized by the wealthy in antiquity; the young women in the foreground of this c. 1470 fresco clearly do not need to darn socks or make their own clothing. Detail from Francesco Della Cossa, *The Triumph of Minerva,* Palazzo Schifanoia, Ferrara, Italy.

pied themselves with embroidery, delicate spinning, illumination of books and reading."[18]

Historian Kay Staniland reports that the church in Britain did not entirely approve of this suspicious popularity of convent embroidery, noting that "as early as AD 747 an attempt was made to check the tendency, clearly already very prevalent in convents, to spend so many hours in needlework." The Clovesho ecclesiastical council in that year urged the nuns to devote more time to religious reading and singing and less to embroidery, but, as Staniland observes, "The problem was not so easily resolved, . . . for throughout the Middle Ages clerics and commentators were to return to the subject of this distraction from the central purpose of convent life."[19] Nuns were forbidden to use embroidery as an excuse to skip Mass, and anchoresses were enjoined by the thirteenth-century *Ancrene Riwle* to "do the plainer kinds of needlework," and admonished: "Do not make purses in order to win friends . . . do not make silk caps or bandages or lace without permission."[20] Knitting gloves for high-ranking male clergy was apparently less problematic, probably because it was considered plain work; textile historian Irena Turnau reports that from the twelfth century forward in Europe, "the majority of bishops' gloves were produced in women's convents."[21]

Since the pleasures of convent needlework were neatly camouflaged as pious zeal for the working of rich ornaments for God's house, the neglect of less enjoyable tasks was grudgingly tolerated by the church hierarchy. Fishing historian Richard Hoffman tells us that male religious in the sixteenth century were similarly counseled against excessive angling when they ought to be busy with their prayers.[22]

Medieval and Protoindustrial Needlework Tools

In part, the spread of embroidery in Europe as a leisure activity in the Middle Ages was made possible by the greater availability of the craft's principal tool, the steel needle. As needlework tool antiquarian Sylvia Groves points out, we have no way of knowing whether the steel needle was known in antiquity in the West, as "no fine needle made of steel could have escaped the destructive action of rust."[23] Bronze needles, commonly found in archaeological digs throughout Europe from ancient times until well into the nineteenth century, are usually too coarse for fine embroidery; many were almost certainly used for the sewing of packs.[24] Bone needles, also common as archaeological artifacts, are often too large for embroidery, as in small sizes bone is a fragile medium.[25] Porcupine quills were used for embroidery in the New World, but the porcupine's Old World relatives did not have the conveniently long defensive adaptation of their pelts that made them prized among Amerindian stitchers.[26]

Silver needles dating from the Carolingian period (eighth century) have been found, but these expensive jeweler-made objects must have been rare to begin with and, like the jeweled needlecases of which a few survive in the British Museum, seem to have been frequent casualties of what Groves calls "the practice, prevalent in those times, of sending outmoded jewelry and silverwork back to the craftsmen, to be broken up and reworked into objects in the prevailing taste."[27]

Iron needles, as economic historian S. R. H. Jones observes, seem to have been available from blacksmiths and wiremakers, but these were, as he puts it, "primitive instruments, sometimes consisting of nothing more than a length of wire pointed at one end and looped at the other." They would thus have been suitable only for sewing coarse, loosely woven cloth or leather that had been prepierced with an awl.[28] The needles mentioned by Italian historian Domenico Sella as having been produced by wiredrawing establishments in sixteenth- and seventeenth-century Spanish Lombardy and distributed by itinerant peddlers are likely to have been of this type.[29]

Until the early to mid-eighteenth century, steel needles were so expensive that they were routinely kept in cases to prevent loss or damage; indeed, in the 1570s the English comedy *Gammer Gurton's Needle* revolved around the loss of a fairly affluent household's one and only needle.[30] Even as late as the early twentieth century, in the southern United States, poverty could reduce households to this preindustrial level of scarcity. Embroidery historian Candace Thurber Wheeler (1827–1923) notes that just after the Civil War, a white Baltimore family known to a friend of hers had only one sewing needle, which was repointed and sharpened regularly by a local jeweler. Among African-Americans and other groups with limited access to economic resources, scarcity persisted longer; novelist Julia Mood Peterkin (1880–1961) described southern black women in 1927 arriving at a quilting party, "each with her needle, ready to sew."[31]

Before the fourteenth century, steel needles in Europe were imported from the Arab world, including Cordoba in Spain, where the making of steel objects that would take both a point and an edge was much further advanced than it was in the early medieval West. John Stow, writing of these instruments much later, in 1598, still called them "Spanish needles," and mentioned a "Negro" steel needlemaker in London in the time of Queen Mary's reign (1553–58); the skilled artisan in question resolutely refused to share his knowledge with the local workers in iron and steel.[32]

The Moorish monopoly on steel needles was, however, broken by Stow's time by the German city of Nuremburg, to which the Islamic world's arts of steel weaponry, armor, and needlemaking had been transferred by the fourteenth century. Prices fell as supplies increased, and by the late 1550s, according to Groves, a "German" (more likely Flemish) needlemaker, Elias Crowse or Krause, had set up shop in England and was busily undermining his former colleagues by teaching British artisans to make steel needles.[33] They learned the lesson well: nearly all of the steel needles used in the United States to this day come from Britain, most of them from the area of Redditch, which was world-famous for its steel needle industry by the eighteenth century. Economic historian S. R. H. Jones describes the technology introduced into England in the 1550s as consisting of "about 20 or so operations" from the drawing of steel wire, which was gauged, cut, straightened, hammered, and punched before tempering, polishing and packeting.[34]

Women who had the leisure to embroider for pleasure bought fine needles of this type, scissors, seam knives (the antecedent to the modern seam ripper), thimbles, and embroidery threads of types that could not be readily produced at home, such as silk floss, at fairs and from peddlers; some may have made use of the trade channels available to the professional tailors, seamsters, and embroiderers in their households.[35] Bronze thimbles and thimble rings had been available since Roman times, and in the Middle Ages the spring shears that survive as modern thread nippers began to be gradually supplanted by scissors of the modern type, including the stork scissor, an elegant device that combined the functions of an awl, stiletto, or bodkin and small thread cutter in a single bird-shaped and highly giftworthy device.[36] The city of Solingen in Germany, created in 1374 from seven smaller municipalities, was one of the first European regions to distinguish itself for cutlery and blade tools; and it remains a major international force in the world of cutlery, including the modern hobby needlework market and the artisan members of the technologically hedonizing Society for Creative Anachronism.[37]

Embroidery frames, being simple works of carpentry, seem to have been locally made until embroidery hoops came into fashion as stretching mechanisms in the eighteenth century.[38] While both professional and amateur embroiderers used frames, scissors, thimbles, needlecases, and, of course, various sizes of needles for different types of work, the tools belonging to noblewomen were personal gear, moved from residence to residence as the family and their entourage devoured the produce of their estates in sequence, and then moved on every few months, literally eating themselves out of house and home.[39] The tools and materials of professional embroiderers remained in their workshops, part of the wardrobe department of the great household, or traveled with itinerant artisans who visited noble households when fitted clothing, mourning clothes, or festival costumes needed to be made.[40]

According to Alexander Neckam, writing of affluent student life in London and Paris in the twelfth century, employers of chambermaids were expected to provide these servants with tools and materials for whatever needlework was required of them, including linen, silk, and gold thread; a leather thimble; scissors; and a variety of needles in sizes ranging from fine embroidery types to the bodkins used for pulling garment drawstrings and ties through their casings.[41] The responsibility of chambermaids for assigned needlework and mending tasks, particularly the latter, remained a tradition in affluent British, European, and American households up through the nineteenth century and passed from a domestic to a commercial context early on: darning and other clothing repairs were among the duties of hotel and tavern chambermaids even in twentieth-century America, a custom which survives in vestigial form in the small sewing kits provided as in-room amenities by modern upscale hotels.[42]

Until well into the nineteenth century, both professional and hobby needleworkers were limited in the number of hours they could work by their access to adequate lighting.[43] Illustrations like those in Diderot's *Encyclopédie* depict embroiderers working in rooms illuminated by daylight shining through what must have been, even in Diderot's time, a very expensive quantity of glass; other depictions and text accounts show needleworkers of all classes and types working outdoors.[44] While candles and firelight were almost certainly used in wealthy households for reading and needlework at night, particularly in the winter months, the former was costly; and both were so far inferior to daylight that in 1303 the provost of Paris, Guillaume de Hangest, prohibited professional embroiderers in that city from working at night.[45]

The Protoindustrial Hobby Needleworker

Medieval gentlewomen, of course, bought their own materials and supplies and chose their own projects, sometimes even designing them themselves. Medieval historian Jennifer Ward notes that "household accounts, like Elizabeth de Burgh's chamber account, refer to the purchase of embroidery thread, and wills occasionally make bequests of items embroidered by the testator."[46] In the early modern period, the most popular hobby projects were purses, book covers, eyeglass cases, potpourri bags, and the newly fashionable pincushions, which had been too insecure as storage devices for anything as precious as a pin in the Middle Ages.[47]

Thread could be purchased at fairs, from peddlers, or from urban merchants.[48] Whether itinerant or not, the retailers of the small wares used by domestic needleworkers were known as mercers or haberdashers in England in the late medieval and early modern period. Their wares included needles for sewing and knitting, pins, thread, scissors, thimbles, and, by 1619, even spectacles for close work. These same merchants also supplied falconers with hoods, bells, imping pins, and other hawking gear.[49]

Until well into the nineteenth century, needlework patterns almost never included text instructions.[50] Early pattern books were of the type familiar to historians of other pre- and protoindustrial technologies: books of illustrations with

Paganino's four methods of transferring an embroidery design to fabric, 1527. *Top left,* tracing a design from a pattern under the fabric, using a candle to create a light table; *top right,* the same technique using light from a window; *bottom left,* scaled redrawing; *bottom right,* pricking and pouncing. The pouncet is visible on the left side of the table. From Alessandro Paganino, *Libro Primo de Recami* (Venice, 1527).

little or no text, which assume that users already know the basic techniques of their craft and are looking for innovations in method and/or design ideas.[51] Thus, any two-dimensional graphic work could be used as a pattern; woodcuts are known to have been used for this purpose after the twelfth century, and book illustrations were routinely copied as well.[52] Embroidery historian Margaret Swain asserts, for example, that a toucan in Mary Queen of Scots' embroidery was copied from a "Byrd of America" depicted in Gesner's 1560 *Icones Animalium.*[53] Alessandro Paganino published illustrations of four different methods of copying embroidery patterns in his 1527 *Libro Primo de Recami.*

Block printing of embroidery patterns developed soon after block printing on textiles appeared in Europe in the thirteenth century.[54] Human figures and curves were and are design challenges for nonprofessional embroiderers, especially in counted-thread embroidery like blackwork, canvaswork and cross-stitch, all of which are worked in a rectilinear fashion on the fabric grid, with curves represented by stepping.[55]

The really well-heeled leisure embroiderers like Mary, Queen of Scots, had their designs painted or drawn for them on stretched fabric, in the manner described by Cennino Cennini in the fifteenth century. This artisan told his readers that "you sometimes have to supply embroiderers with designs of various sorts. And for this, get these masters to put cloth or fine silk on stretchers for you, good and taut."[56] Cennini was clearly referring to professional embroiderers, not noble hobbyists, but the technique was the same. Swain reports itinerant artists who drew embroidery patterns for affluent clients; those at the English and most other royal courts were professionals.[57] In fifteenth-century Florence, the "broderers" bought designs from the artists who drew cartoons for weavers, the same drawings being adaptively reused many times in various media.[58] Textile curator Marie-Anne Privat-Savigny has identified a number of European artists in the period 1570–1610 who painted or drew embroidery designs for royal patrons and anyone else willing to pay for them.[59]

When needlework pattern sheets began to be published in Germany in the late fifteenth century and pattern books in 1523, they too were mined by artisans of many media looking for good representations of figures, foliage, flowers, borders, and other subjects difficult for artisans without drafting training to draw for themselves. Many of these patterns were represented on point paper, so that users could more easily work out the difficult stepped curves and/or rescale them onto a larger or smaller grid.[60] While nearly all lace in the early modern period was the work of professionals, the germ of a lacemaking hobby movement had begun among the aristocracy, especially those in the more comfortable of the needlework-oriented convents, and published patterns provided convenient sources of lace designs.[61] The crossover from embroidery to lace in the form of drawnwork and cutwork appears to have been popular in the sixteenth century; an Italian designer devoted an entire pattern book to it in 1587.[62] His work has since been reprinted many times. A large number of samplers in museum collections document this crossover; Marcus Huish shows several examples in his *Samplers and Tapestry Embroideries*, originally published in 1900, including one seventeenth-century example from colonial America.[63]

While samplers may predate the printed pattern book, it is worth noting that nearly all of those that survive are from the era of published needlework patterns.

Printed patterns were, of course, more expensive than samplers and did not include instructions for working.[64] A sampler—"exemplar"—embodied knowledge both of design and of technique and could be used as a reference for both. Moreover, samplers, of which one of the earliest surviving dated examples is marked 1598, could be exchanged between needleworkers, even internationally, transmitting technical information across both time and distance.[65] Kay Staniland says of samplers that although they "are known only from the early sixteenth century . . . they must certainly have been worked for quite some time before this." She regards them as a "manifestation of the rise of the amateur needlewoman," pointing out that many were made by adult women as technical documentation, not as pedagogical exercises. She goes on to say that "in great households, indeed, samplers seem to have been collected together into reference libraries: an inventory of Joan the Mad, Queen of Spain, dated 1509, lists no fewer than fifty samplers, some worked in silk, others in gold thread."[66] Darning samplers were, of course, much plainer, and were probably not much used by noblewomen themselves, although they may have used them to teach darning stitches to their servants.[67]

British antiquarian Averil Colby quotes Barnabe Riche's 1581 work *Of Phylotus and Emilia*, which makes clear the role of the sampler as an embodiment of tacit knowledge for affluent women with leisure to spend on needlework: "Now, when she had dined, then she might go seke out her exemplars, and to peruse which worke would doe beste in a ruffe, whiche in a gorget, whiche in a sleeve, which in a quaife, which in a caule, whiche in a handcarcheefe; what lace would doe best to edge it, what seame, what stitche, what cutt, what garde: and to sitte her doune and take it forthe by little and little, and thus with her needle to passe the after noone with devising of things for her owne wearynge."[68]

Richard Brathwait's virtuous woman of 1631 reads improving books, prays, sews, and works her sampler: "Some books shee reads, and those powerfull to stirre up devotion and fervour to prayer; others she reads, and those usefull for direction of her household affairs . . . She is no busie-body, nor was ever, unless it were about her family, needle or sampler." Brathwait seems to consider the reading of household manuals as improving as prayer.[69]

Between the emergence of samplers and that of pattern books in the early modern period, it is possible to discern a trend away from the narrated-demonstration tradition of transmitting technical information from one artisan to another, toward a democratizing trend that allowed information exchange without the in-person mediation of professionals or the expense of travel to a knowledgeable artisan.[70] This was, of course, part of a larger trend in which print broke through temporal, geographic, and class boundaries in a broad range of activities in early modern Europe and elsewhere, with treatises published on fishing, falconry, warfare, architecture, cooking, sex, farming, milling, travel, gardening, mining, farriery, polite behavior, and a host of other subjects for which there was an apparently insatiable demand from inquiring minds.[71] For the users of many of these works, including the needlework pattern books, literacy was not an issue, nor was the vernacular language of the country of publication, because much of the substance was in the illustrations. Some needlework pattern books, including that of Giovanni Battista Ciotti in 1596, explicitly identified hobby needle-workers as their market; the London publisher announced that Ciotti's *Booke of Curious and Strange Inventions* was printed "for the profit and delight of the gentlewomen of England."[72]

Staniland, quoted above, seems to be arguing, as I have done here, that the hobby needleworker had her origins in the Middle Ages, but this view is at odds with the views of leisure theorists and our fellow historians regarding the early appearance of hobbies in Western culture. As we have seen, Steven Gelber asserts that hobbies did not exist until the nineteenth century. British historian Peter Burke, however, argues against a sharp break between preindustrial and industrial-period leisure, asserting that "leisure activities of different kinds, whether for children or adults, males or females, came to be viewed as less and less marginal from the late Middle Ages onwards."[73] Burke is arguing against an essentially Foucauldian model that distinguishes between a preindustrial concept of leisure as "festival culture" that makes "the emergence of leisure . . . part of the process of modernization." I concur with Burke that this "'rupture' between the two periods" is a figment of the constructivist imagination, but take issue with his idea of the marginalization of leisure in the preindustrial period. Surely Burke does not intend us to believe that members of the European aristocracy were "marginal," although they certainly represented a tiny minority. A better term to describe the spread of leisure activities is *democratization*—the time, materials, and tools became available to a larger number of persons. Burke also notes, correctly, that this process of change included the precipitous rise in the number of the how-to manuals that we have already discussed. Needlework, as we shall see, was only one of many hobbies and leisure activities that benefited from the rise of print.

Print, Pleasure, and Protoindustrial Leisure Artisans

Thomas Jefferson once said of himself that, although he was by then a very old man, he was still a very young gardener.[74] No one in that age or any other could live long enough to understand everything that happens on even a very small plot of cultivated ground.[75] For the hobby gardener, this recognition of knowledge too vast to be encompassed in one lifetime is a major component of the allure, as it is in all hobbies: there is always something more to be learned.

The lust for knowledge familiar to scholars illuminates much of the literature of passionate hobbies and leisure activities, as well as of professional pursuits like that of historiography. For the needleworker, woodworker, or ceramicist, there was and always is some technique not yet mastered; for the collector, some object not acquired; for the gardener, some species not yet successfully brought into cultivation.[76] Print culture broke the chain that fastened the transfer of artisanship to narrated demonstration and to formal or informal apprenticeship, and allowed literate learners to consult the experience of a broader range of practitioners, who might be at great temporal and spatial distances. It permitted ordinary persons to consult authorities whose personal tutelage would have been prohibitively expensive and permitted the replication of experiments in artisanship with the assurance that at least some of them had succeeded before. As William Eamon observes in his magisterial work on how-to books of the early modern period, *Science and the Secrets of Nature*, "We trust recipes because we know that behind them stands someone who does not use them."[77]

What the early modern period brought to hedonizing technologies was not only a broadening of the base of practitioners, but a second derivative of hobby growth: the passion to collect information about one's primary leisure enthusiasm. Pattern books, erotica, manuals, recipe books, technical treatises, guides, catalogs, inventories of collections, and every kind of documentation familiar to modern hobbyists had their popular origins in the infancy of print, when the technology of reproducing texts and illustrations created a market of grown children run amok in an informational candy store. Hobby technologies have also a third derivative: their institutionalization in museum exhibitions and collections, which entails their intellectualization as subjects of antiquarian research, a development that for embroidery and lace in Britain and the United States was well under way by the late nineteenth century.[78]

We have already noted how, in the case of needlework, some of the early published pattern books addressed hobbyist "gentlewomen," and how these works focused on the two genres that had been hedonized up to that time: embroidery and some kinds of lace, especially those made with a needle.[79] The bibliometric evidence—the number of published titles—in fact supports the hypothesis that only these two among the needle arts had moved far enough along the hobby-technology path to have created a significant leisure market for patterns, tools, and materials. [80] This evidence is set forth in more detail in Appendix B. Biblio-

The second derivative: from restoring antique autos to collecting automobilia. An exhibit in the Piccadilly Museum of Transportation, Butte, Montana, 2007. Photo by John B. Northup, reprinted by permission.

metrics is underutilized by historians, but even used in the relatively simple-minded way it is applied here, it can reveal relevant patterns of interaction among texts, their producers, and their consumers.[81] The bibliometric methodology used in this book is explained in the appendix.

As we move into the industrial era in the next two chapters, we will see dramatic changes in this picture, as the hobby markets for knitting, crochet, tatting, quiltmaking, sewing, and dressmaking expand to support large literatures of patterns, instruction books, and magazine titles. In one genre of needlework, however, we will see little change in the bibliometric situation: in nearly five centuries of needlework publication, few works in English appear in WorldCat on the subject of darning, and one is the 1942 *Make and Mend for Victory*.[82] None of the titles mention pleasure or delight. Darning has yet to make a full transition from work to pleasure, although its tools have recently come into vogue as collectibles, and there are many examples of darning samplers in museum collections.[83]

Other leisure activities and hobbies benefited from the print revolution that began in the late fifteenth century, including hunting, hawking, fishing, and gardening. Among hobby crafts, however, the growth of published literature in the protoindustrial period is significant, as are uses of the rhetoric of pleasure in the marketing of these documents. Home preserving, for example, acquired by the mid-seventeenth century a literature of instruction books and recipes for hobbyists, as well as an instruction literature for those performing the same tasks as work. *A Queen's Delight; or, The Art of Preserving, Conserving, and Candying*, first published in 1658, went through ten editions by 1698.[84] The publishers of *The Queen's Closet Opened* claimed in 1662 that the recipes contained therein "were presented unto the queen by the most experienced persons of the times, many whereof were had in esteem, when she pleased to descend to private recreations."[85] While it is, in fact, difficult to imagine Elizabeth I in an apron, stirring a kettle of strawberry preserves with a wooden spoon, it is significant that the activity of preserving was apparently regarded as a suitable royal recreation. Luis Vives, in Richard Hyrd's 1529 translation of the Latin work known as *Instruction of a Christian Woman*, endorses such activities for noblewomen, asserting that "those who say that sewing, cooking, weaving and spinning are base exercises performed because of poverty and not out of fitness to the sex, are merely ignorant, for, setting all other reasons aside, noble, rich, and great women must occupy themselves in such matter to avoid idleness."[86] Here, as one might expect, the argument is virtue, not pleasure.

The publisher of Hannah Woolley's *The Accomplish'd Lady's Delight in Preserving, Physick, Beautifying, and Cookery*, dated 1675, however, uses the rhetoric of pleasure and noble leisure to interest the reader in the book's contents, as does Elizabeth Grey, Countess of Kent, in her 1653 *A True Gentlewoman's Delight wherein is contained all Manner of Cookery; Together with Preserving, Conserving, Drying and Candying*, which her publisher goes on to commend as "very necessary for all ladies and gentlewomen."[87] We see in these titles the beginnings of hobby cookery more generally, of which I shall have more to say later on.

The literature on home preserving contrasts with contemporaneous sources that were clearly written for and marketed to persons, mainly women, for whom preserving was a task—possibly a pleasant task, but a requirement rather than a recreational activity.[88] Their title pages are generally bare of references to pleasure or recreation, although in some cases the works were written by the same authors as those who addressed the hobby market, such as Hannah Woolley.[89]

As the reference to "Physik" in Woolley's title suggests, part of the work of preserving on estates was the making of home remedies for winter storage in the still-room. Conquest, slavery, and trade with the New World brought hitherto-unknown quantities of sugar into British and European households at accessible prices, and the confectioner's art was achieving culinary glories undreamed of in medieval kitchens.[90] Both the ability to stock and manage a still-room and the skills of confectioners were highly prized in households affluent enough to be able to employ these artisans or to buy their products; householders who were not affluent could endeavor to learn the skills themselves from one of the less lofty instruction books published in the seventeenth and eighteenth centuries. The literature of preserving-as-work includes both guides for servants aspiring to good positions in wealthy households and manuals for "ladies and gentle-women" to do the work themselves or to supervise its performance by servants.[91] The duality of this literature of preserving persists even today, as we shall see in the last chapter.

The Hedonizing Landscape: Gardening, Shooting, Brewing, and Breeding

The print revolution in guides to leisure activities that began in the sixteenth century, and the general expansion and democratization of leisure that began in the early modern period, were, unsurprisingly, nearly always gendered in expression. Even when the same leisure activity, such as gardening, could be and was engaged in by both sexes, there were gender barriers between the hobby and professional sides of the craft. In the case of gardening, men were generally the professionals who earned their living in landscape design and maintenance; women and men of leisure were the hobbyists who cultivated flowers and exotic fruits and vegetables for pleasure.[92] Clergymen were encouraged to devote their leisure hours to the refined, innocent, and artistic joys of gardening. For example, a pomological manual published in England in 1672 declared itself to be *A Short and Sure Guid* [sic] *in the Practice of Raising and Ordering of Fruit-trees: Being the Many Years Recreation and Experience of Francis Drope, Bachelour in Divinity, late fellow of Magdalen Colledge in Oxford.*[93]

The literature of gardening published before 1800 includes more than thirteen hundred book titles and three serial publications. The scientific literature on plants and husbandry was often in Latin at this period, but guides for hobbyists and dabblers were in the vernacular and, like the literatures of needlework and hobby preserving, addressed their readerships in the rhetoric of pleasure. Leonard Meager, for example, sounds remarkably modern in his 1688 *English Gardner*, when he tells prospective readers that his book is "fitted for the use of all such as delight in gardening, whereby the meanest capacity need not doubt of success," making it, in effect, a seventeenth-century antecedent of *Gardening for Dummies*.[94]

Title pages of seventeenth- and eighteenth-century gardening manuals usually emphasized the joy, virtue, aesthetic rewards, and challenges of hobby gardening. In Britain and the United States flower gardening was considered an appropriate hobby for women of leisure, most of whom were expected to employ professional gardeners for the heavy work of digging, planting, and weeding, reserving for themselves the more enjoyable tasks such as selecting species and varieties, planning and arrangement of plots, cutting blooms for the house, and sometimes deadheading at the end of the season. Works intended for this audience were excruciatingly genteel in tone, like John Parkinson's 1656 *Paradisi in Sole Paradisus Terrestris*, for those who yearned for "a choise garden of all sorts of rarest flowers"; William Hughes' 1671 work, *The Flower-Garden*; and Samuel Gilbert's *Florist's Vade Mecum* of 1683.[95] Some authors offered their readers detailed instructions for such death-defying horticultural and pomological stunts as raising pineapples in Scotland and oranges in Holland, achievements which could have been of very little, if any, economic utility, but must have afforded many hours of the kind of pleasure Rex Stout depicts in Nero Wolfe's fictional orchid-raising in New York City.[96]

Gardeners of more robust sensibilities and tastes than the "florists" and genteel amateur pomologists had their own literature of hobby and leisure gardening. Thomas Barker, for example, wrote on gardening in his 1654 *Country-Man's Recreation*, with sections on raising hops and fishing along with "the art of planting, graffing, and gardening."[97] Matthew Stevenson, writing in 1661, set out to fill every gentleman's life with a full year's leisure, offering a wealth of seasonal "labour" and diversions, including "plowing, sowing, gardening, planting, transplanting . . . as also of recreations, as hunting, hawking, fishing, fowling, coursing, cockfighting: to which likewise is added a necessary advice touching physick . . . : lastly, every moneth is shut up with an epigrame: with the fairs of every month."[98] For gentlemen with catholic tastes in leisure activities, useful or otherwise, a 1684 title by one "S.J." promised *Profit and Pleasure United*. Subtitled *The Husbandman's Magazine*, this volume claimed to be the most exact treatise of horses, mares, colts, bulls, oxen, cows, calves, sheep, swine, goats, and all other domestick cattle serviceable, profitable, or usefull to man . . . together with easie and plain rules and methods for improving arrable and pasture-lands and the like, improving most sorts of grain to the best advantage, and what is necessary to be observed in sowing and harvesting, the management, improvement and preservation of fruit-trees, plants and flowers, the manner of ordering flax, hemp, safforn and licrish, with directions for the increacing and preserving of bees, and many other things of the like nature: to which is added the art of angling, hunting, hawking, and the noble recreation of ringing, and making fireworks: the whole elusterated [sic] with copper cuts.[99]

Joseph Blagrave promised his readers in 1670 a "gentlemans heroick exercise, discoursing of horses" as well as horticulture, both ornamental and economic. The sections on farriery in this work, however, address the subject as work rather than as recreation.[100]

It was not unusual for works on gardening for gentlemen to include guides for bell-ringing, a distinctively British leisure pastime, as well as more virile recreations like cockfighting, coursing, hunting, shooting with a crossbow and/ or firearm, bowling, tennis, and even billiards.[101] S.J.'s inclusion of fireworks in a manual for leisure hobbies was by no means unique: John White's 1680 *Art of Ringing*, a manual for hobby bell-ringers, included "artificial fireworks; being directions to order and make moulds . . . rockets, fisgigs, and other curiosities too tedious to mention" along with gardening instructions.[102] Nathaniel Nye also mentions fireworks in his famous *Art of Gunnery* of 1670, declaring his intention to familiarize the reader with their use in "war and recreation."[103]

Among the earliest documented masculine hobby crafts in Britain and America is home brewing, an art that had historically been a household task associated primarily, though not exclusively, with women.[104] In the seventeenth and eighteenth centuries, a number of guides to home brewing were published in Britain; one appeared in Philadelphia in 1805. Gervase Markham, a veritable Renaissance man of seventeenth-century recreation manuals for gentleman, offered a cornucopia of information for the man of leisure in the late seventeenth century, including one of the English language's earliest guides to home brewing as a hobby.[105] His work also included the more traditional instructions for gardening, hunting, shooting, ringing, fishing, care of horses, and agricultural pest control.

The capture and breeding of songbirds as a hobby also emerged in the eighteenth century, documented and described in works such as George Wright's

Hawking c. 1600. While in theory, hawking and falconry were forms of hunting, the seventeenth-century author Richard Braithwaite estimated that the costs of falconry were more than eight times the value of the quarry taken by it. Note the rich dress of the participants. ClipArt.com.

1728 *Bird-Fancier's Recreation,* creating a less costly and space-intensive form of leisure husbandry than the gentlemanly, largely rural, arts of breeding falcons, dogs, and horses, all of which were mainly "off-season" adjuncts of the sports of hunting, hawking, and shooting.[106] By the mid-nineteenth century the raising of small birds was recommended as an entertaining hobby for young boys.[107]

Close on the heels of these developments, and under the same protoindustrial and technological influences, artisan crafts-as-work underwent a two-century revolution in production techniques in which textiles were one of the vanguard industries. The knitting of stockings was mechanized by William Lee in 1589, although it was forty years before the industry accepted the innovation, and John

Kay patented the fly shuttle for weaving in 1730.[108] While the change was gradual, with many bottlenecks and lags in the flow of technological change through textile artisanship to factory production, each decade of industrialization in the eighteenth and nineteenth centuries moved the processes of commercial textile production and ornamentation farther from those of leisure artisanship and moved new techniques into the economic space occupied by the hedonizing technologies of textile hobbies.[109]

CHAPTER THREE

Hedonization and Industrialization

Diverging Paths in the Eighteenth
and Nineteenth Centuries

Nerylla Taunton's lavishly illustrated guide for collectors, *Antique Needlework Tools and Embroideries,* opens with a chapter on the seventeenth century, because, the author tells us, "apart from thimbles" this century is the one from which "the earliest needlework implements [are] available today for collectors."[1] Although Taunton does not say so, the abundance of these collectibles is at least in part an artifact of the increase in supply of these tools, and of their materials as well, in the early modern period. Survival of these artifacts is, of course, subject to the biases of collecting and connoisseurship described in Appendix A.

While I have postulated that some medieval and early modern embroiderers were hobbyists in the modern sense, from the late seventeenth century forward, consumers' love of tools and materials for leisure activities and hobby artisanship is clearly and unambiguously documented. The existence of hobby craft in textiles is no longer speculative or conjectural. In this chapter and the next two, we will see the demand for pleasure-producing tools and materials expand through the middle class in the nineteenth century and the entire industrialized world in the twentieth. We will see parallel developments in other kinds of hobby crafts, and in other tool-based leisure activities, such as woodworking, shooting, cooking, ham radio, and ceramics.

During the period that in the United States is called the colonial and early federal, English and French fashions in embroidery and lace gradually adapted to transplantation in the New World. Most materials, and nearly all tools except embroidery frames, were imported and expensive. In Europe, two new needle arts were evolving out of older traditions: crochet, in all probability from the hooks used in tambour embroidery, and tatting, most likely from the shuttles used for netting. The connection of both with commercial production was very limited and very brief; both became leisure arts for the affluent soon after they appeared on the Continent and in Britain, and were democratized in the United States and

Britain within a few decades of the invention of smooth, plied, hard-twisted cotton thread for sewing and needlework in the first quarter of the nineteenth century.[2]

Industrialization in the Workbasket: Embroidery, Lace, Crochet, and Tatting

It is significant to America's history of leisure textile artisanry that much of the European colonization of North America took place in the early modern period, when tools, materials, and patterns were beginning to be widely available. By the middle of the seventeenth century thimbles, scissors, pins and needles in rust-preventive oiled paper packets were being shipped into Virginia and Massachusetts in large quantities; cutlers and jewelers in urban areas were producing needles, pins, thimbles, and scissors; and fancy needlework was being produced by the wives and daughters of the prosperous and leisured few.[3]

By the early eighteenth century, American and British town dwellers could buy materials for needlework in shops or from peddlers; those in the country bought from peddlers, sent orders to towns, or ordered directly from importers.[4] In America, as Patsy Jefferson told her father, Thomas, in 1787, it could be very difficult to get "proper silks," a condition that was soon to be remedied.[5] In New York City by 1800 jewelers sold embroidery and lace patterns, thimbles, needles, tambour hooks, netting tools, and even "ladies' knives," the hand-crafted equivalent of modern seam rippers. Some jewelers also made and sold fishhooks to both food-oriented and hobby anglers.[6]

Embroidery frames, as in Britain, were made by local carpenters and furniture-makers. The workboxes of the affluent contained not only thread, needles, thimbles, and scissors but often vinaigrettes for drying the hands, mirrors, combs, hairpins, and porcelain or the bone or metal needlecases called etuis.[7] Various kinds of ink, ground nutmeg, and fine soot were sold for pouncing embroidery designs through paper onto fabric. Faces, which were very difficult to render in embroidery, were often painted or inked onto the fabric by professionals. The antiquarians Harold Eberlein and Abbot McClure, commenting on American embroidered pictures of the eighteenth century, remark that "the drawing was usually creditable, and this is to be explained by the fact that, in many instances, ready prepared designs, kept in stock by the purveyors of embroidery materials, were traced or pounced upon the satin, silk or bolting cloth, all ready to be filled in with stitch-work."[8]

Another antiquarian, George Francis Dow, adduces evidence on this point from advertisements printed in New England newspapers between 1704 and

1775, noting a "counterpane stamper" and a "linen stamper" in 1747, and a "japanner" who advertised drawing on satin or canvas for embroidery in 1758. Dow also lists a Boston widow in 1737 who would either draw embroidery for others, or work it herself, a traditional combination of services for a professional embroiderer, and another Bostonian, daughter of a deceased embroidery designer, who advertised the sale of her mother's patterns in 1747. The city was evidently well supplied with embroidery services, as one merchant provided "yellow canvas and canvas pictures ready drawn, a quantity of worsted slacks in shades" in 1757, and another the "drawing and working of Twilights" in 1775.[9] A Mrs. Whitby of Philadelphia, advertised in 1759 that she taught people "to pickel [*sic*], preserve and make fine Paste" and also taught "all sorts of embroidery in gold, silver, silk or worsted."[10] Clearly, the leisure market for tools and materials was well launched.

Silk thread was, of course, an imported article, and hard-twist cotton thread had yet to be invented, although soft cotton yarn was available for knitting and candlewicking, a type of raised (three-dimensional) embroidery. Cotton floss was available in some places for embroidery as well. Patterns were still drawings only, without instructions; British magazines for women, which were available in at least the urban areas of the American colonies, published patterns for embroidery as early as 1776.[11]

For ordinary sewing, handspun linen thread, usually made from plied yarns spun from the short, soft fibers called tow, could be bought from the small wares sellers called haberdashers (mentioned in Chapter 2) or from peddlers. In rural areas, linen thread was made at home, but we have no reason to think this practice was common in cities; certainly Elizabeth Drinker, in her Philadelphia diaries spanning the years 1758 to 1807, does not mention home spinning of thread, although many other textile activities are described, including both plain sewing and embroidery.[12] Sam Bass Warner reports that Philadelphia's Middle Ward had a "threadmaker" in the eighteenth century; it is likely that most thread was purchased from professionals rather than made at home, at least in urban areas.[13]

As a respectable use of aristocratic leisure, embroidery and lace-making were as much the mainstays of hobbies among affluent women in colonial and early federal America as they had been in the Old World in the medieval and early modern periods. The popular types of leisure needlework in the late colonial and early federal periods were crewelwork, canvas embroidery (needlepoint), cross-stitch in various forms, tambour work, and a variety of methods of "flowering"— that is, embroidering flowers in silk and chenille yarns and threads. Although some girls and women have recorded for us that they preferred other activities, such as reading, to needlework, there were occasions when only needlework was acceptable, as Thomas Jefferson advised his daughter Patsy in 1787: "In the country life of America there are many moments when a woman can have recourse to nothing but her needle for employment. In a dull company, and in dull weather, for instance, it is ill-manners to read, [and] it is ill-manners to leave them."[14]

The novelist and evangelical author Hannah More (1745–1833), writing in *The Lady's Pocket Library* in 1792, concurred with Jefferson, asserting that "ornamental Accomplishments," like needlework and music, "but indifferently qualify a woman to perform the *duties* of life, though it is highly proper she should possess them," presumably for purposes of enjoying leisure hobbies, whether in "dull company" or not.[15] Sarah Smith Emery (1787–1879) wrote a charming autobiography, *Reminiscences of a Nonagenarian*, in 1879, in which she recalls her teacher of some three-quarters of a century before as "a most accomplished needlewoman, inducting her pupils into mysteries of ornamental marking and embroidery. This fancy work opened a new world of delight."[16] Nancy Shippen Livingston as well, seems to have enjoyed the "Accomplishments," including needlework, that Jefferson and More endorsed; in 1784, she budgeted five hours a day for them, more time than she spent at meals and more than twice as much time as she allocated to managing her household.[17] A few decades later, after the founding of America's most famous nineteenth-century women's magazine, *Godey's*, an article on embroidery opined that "numerous as are the subjects treated in this work, there are few which furnish a more pleasing occupation than Embroidery . . . It . . . offers a graceful occupation, and an inexhaustible source of laudable and innocent amusement."[18]

The rudiments of an education in plain needlework, including seams, hemming, darning, and sometimes knitting, were recognized as a necessity for all girls and women in eighteenth- and early-nineteenth-century America, and sometimes also for boys of low status, such as orphans, who might find themselves in occupational situations—as sailors, loggers, soldiers, or farm or ranch hands—in which it was unlikely that a woman would be always available to mend their clothing and bedding.[19] Darning samplers attest to the technical competence expected even of these "plain work" students.[20]

"Marking"—what modern Americans now call "cross-stitch"—was usually taught to both plain-work and fancywork students and could be used for either type of work, depending on whether the artisan was marking her employer's linen and small clothes (underwear, collars, and cuffs) to identify them at the laundry, or stitching decorative objects as a hobby.[21] On the plain-work use of the stitch, and its appearance on samplers, Marcus Huish shrewdly pointed out that "a sampler would seldom, if ever, be used as a text-book for children to learn let-

ters or figures from, except with the needle, and the need for lettering and figur-
ing upon them would, therefore . . . only arise when garments or napery became
sufficiently common and numerous to need marking."[22] Huish would not have
known the word "protoindustrial," but he correctly identifies the appearance of
"marking" stitch in seventeenth-century samplers as an indication of increasing
textile abundance. He so wholeheartedly approved of plain work as an occupation
for the juvenile *hoi polloi* even of his own early twentieth century that he objected
to their apparent preference for reading: "The decay of needlework amongst the
children of the middle classes may perhaps be counterbalanced by other useful
employments, but undoubtedly with those of a lower stratum of society the lack
of it has simply resulted in their filling the blank with the perusal of a cheap literature, productive of nothing that is beneficial either to mind or body."[23] One
shudders to imagine what Huish would have thought of television.

Eighteenth- and nineteenth-century schools, workhouses, and orphanages
taught darning, mending, sewing, and marking, sometimes along with the three
R's.[24] Instruction in fancy needlework (embroidery and lace) was reserved for the
more expensive academies and schools, which taught them along with music
and French. The needlework of necessity is quite clearly distinguished from
hobby embroidery and lacemaking in the advertisements of schools that operated
in Pennsylvania, North Carolina, and New York in the late colonial and early federal periods, with some schools offering fancy work, music, French, and drawing
as add-ons at additional cost.[25] These latter were the kinds of skills intended as
genteel amusements to beguile the leisure hours of women who had assistance
with less glamorous needlework tasks like knitting socks, making and mending
clothing, and hemming linens.

In Philadelphia in 1762, for example, a girl under twelve years old could receive instruction for two hours a day in "Berlin or Dresden Needle-Work . . . for
Ten Shillings Entrance, and Thirty Shillings a Quarter." At the Young Ladies'
Academy in the same city in 1803, parents paid a pricey seventy pounds a year for
tuition and board, with music and embroidery billed separately at five pounds,
one shilling, threepence, and two pounds, five shillings, respectively, payable per
quarter. The embroidery course was the same price as laundry service for the quarter. Students at the Bethlehem (Pennsylvania) Boarding School in 1787 paid extra
quarterly fees of seventeen shillings, sixpence, to learn the refined arts of drawing
and tambour embroidery. [26] The Westtown school, run by the Quakers in Pennsylvania, remains famous for the high standard of needlework taught there.[27]

Elaine Forman Crane, editor of the 1991 edition of Elizabeth Drinker's diaries
covering the years 1758 to 1807, observes of Drinker that "her skills . . . included

needlework, from which she derived great pleasure throughout her life."[28] As Crane points out, Drinker, of a prosperous Quaker family in eighteenth-century Philadelphia, did both plain- and fancywork, the former often with outside help. Her diary notes the occasions when she sallies forth to buy materials for fancywork, obtaining her "cruels" (crewels) from a retailer named Smith and her "Nitting-Needles" from "Preist's." As Drinker's children were born and her responsibilities multiplied over the years, needlework references were usually eclipsed by detailed accounts of the deranged workings of her family members' digestive tracts, and sad tales of death and illness in the community, to say nothing of the disturbances of the American Revolution.[29] Significantly, as we shall see, she notes knitting with silk as a leisure pastime later in her life; this hobby was to gain popularity gradually in the decades after her death.

The received wisdom about the origins of crochet is that the art arose from experimentation with the hooks used for tambour embroidery after 1764. Tambour was the first type of embroidery known to have been stretched on a hoop; British historian Therle Hughes describes the technology thusly: "In a tambour frame two hoops fit inside each other with the fabric between them, a rim of cork or felt to the inner hoop, or a binding of ribbon, ensuring a firm grip,"[30] The chain stitches used in eighteenth-century tambour work could be worked "in the air," as needleworkers say, looped into each other to form a fabric.[31] So far all efforts to locate the invention of crochet more precisely in time and place have been fruitless. Lis Paludan, whose research on this subject is the most meticulous to date, could find no examples of crochet in the United States, the British Isles, Asia Minor, or Europe that predated the early nineteenth century, and these early examples were not of the lacy material familiar to modern needleworkers but the very pedestrian single-crochet technique known as "shepherd's knitting." Even in the Irish and British convents, where the "consensus history" of needlework had located the origins of crochet in the West, Paludan found no work earlier than the mid-nineteenth century. Because at least one type of crochet, so-called "Tunisian" crochet, was believed to have originated in Asia Minor or perhaps farther east, Paludan searched repositories in the Middle East and North Africa, but found no crocheted fabrics before the 1920s.[32] British textile historian Pauline Turner also found no European references to crocheting before the middle of the nineteenth century. She documents the use of crocheted stitch exemplars instead of printed patterns as late as 1900 in Britain.[33]

Tatting—an apparent hybrid of netting and knotting, using the same hooks as those used for crochet—has equally mysterious origins, although by tradition, its birth is associated with eighteenth-century Italy, where it was known as *occhi*,

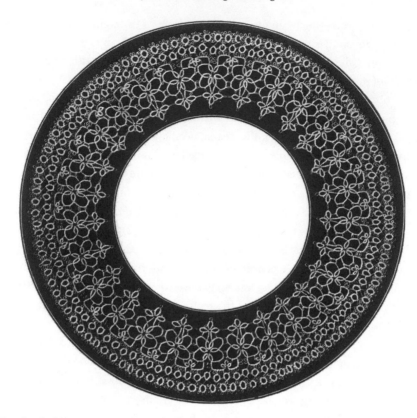

Tatted yoke by Georgette Batt, 1916. Tatting is by definition a decorative technique.
By the nineteenth century it retained only vestiges of its origins in knot-making.
From *The Priscilla Yoke Book* (Boston, MA. Priscilla Publishing, 1916).

"eyes," for the circular formation of its delicate picots and loops. Both netting and
knotting were worked with silk and linen thread in the eighteenth century, becoming inexpensive, popular hobby arts in the nineteenth, after the beginning of
industrial production of cotton thread that was suitably smooth and strong at the
tensions required.[34]

Hedonization Holdouts: Mending, Plain Sewing, Knitting, and Bobbin Lace

Charlotte Brontë's 1847 novel, *Jane Eyre*, more often read for its gripping narrative and irresistible romance than for its needlework references, offers a perfectly
orthodox view of the needlework social hierarchy in mid-nineteenth-century

Britain. Brontë's female characters consistently engage in the textile activities most appropriate to their station. Jane herself, penniless but brought up and educated among the relatively genteel, can mend, darn, and knit, but also has ladylike accomplishments: beadwork netting with silk, "work on muslin and canvas," and painting in watercolors. After leaving Thornfield, she teaches what she later calls "the finer kinds of needlework," in the village school at Morton.

In the prosperous Reed household, where Jane spends part of her childhood, a servant named Bessie irons, mends, and tells stories to the children. Later, at Lowood School, the young women learn to cut and sew their own clothing and thus are prepared to become dressmakers, a qualification by which Jane later attempts (unsuccessfully) to earn a living. The intended practicality of the schoolgirls' needlework education is suggested by the "brown Holland" linen pockets of the apron type that they wear on the front of their dresses, "destined to serve the purpose of a work-bag." The honest and kindly housekeeper of Thornfield, Mrs. Fairfax, is almost invariably depicted with knitting in her hands; Brontë does not tell us what she is knitting, although there is a somewhat bitter reference to stockings in Chapter 12. The sturdy but bibulous wife-keeper of Thornfield, Grace Poole, lives up to her seamstress camouflage in Chapter 16. Another kindhearted servant, Hannah, the elderly handmaiden of the ladylike but impoverished sisters at Moor House, where Jane seeks shelter after her flight from Rochester, knits, brews, bakes, washes and irons.[35] Brontë writes of all these plain- and fancywork activities with an easy familiarity, clearly confident that readers will recognize them for the social markers they are.

Brontë's account is consistent with the tradition that mending and darning are the most tedious drudgery of the needle crafts, penitential exercises to be gladly handed over to subordinates as soon as resources permit. Mending and darning—historically the province of serving women, wives, mothers, and sisters of the none-too-affluent; of female children; and of unmarried women living in larger family households—were mandatory parts of female education in most times and places before the twentieth century. Even wealthy women had to learn the basic techniques so as to be able to instruct and supervise their children and servants. Darning, the stepchild of both embroidery and knitting, requires an intimate knowledge of both woven and knitted fabric structures, since in its most sophisticated form, the mender is expected to reproduce the original fabric in structure, texture, weight, and color. While little artistic imagination is required, the technical obstacles are formidable, and unlike, for example, embroidery, the finished product is not some new item in which one may take pride and show off with pleasure, but a refurbished worn object that advertises the mender's inability to buy or make a new one, a veritable badge of poverty. Wilmet, the eldest daughter in the thirteen-member orphaned and economically stressed Underwood family in Charlotte Yonge's 1873 novel *The Pillars of the House*, expresses this ambivalence about darned objects when her elder brother complains of the shabbiness of their floor covering:

> "That carpet is solid darn," said Felix. "We tried one evening, and found that though the pattern of rose-leaves is a tradition, no one younger than Clem could remember having seen either design or color."
>
> "You should not laugh at it, Felix," said Wilmet, a little hurt; for indeed her mother's needle and her own were too well acquainted with the carpet for her to like to hear it contemned.[36]

In fiction and in the instructional guides for women so popular in the nineteenth century, mending and darning are typically portrayed as inherently virtuous but unpleasant activities that must be endured, preferably cheerfully. The virtue did not, however, entirely eclipse the stigma of poverty; both the virtue and the stigma associated with plain work are vividly evoked in Louisa May Alcott's 1869 *Little Women*. Only the charmingly impractical Amy knows how to crochet; all the other March sisters are dutiful laborers in the vineyard of plain work. By contrast, Jane Austen's Elizabeth Bennett and her sisters make themselves quietly attractive in company with their fancy needlework, and it is their servant Sally who does the mending.[37] Female fictional characters in nineteenth-century novels are not infrequently depicted hiding their mending baskets upon the arrival of gentlemen callers, and hurriedly whisking some piece of decorative fancywork into their laps.[38] Others, both male and female, are complimented as fictional pillars of rectitude because of their devotion to the virtues of plain work, or criticized for their frivolous interests in embroidery, lace, or netting with silk. The truly depraved did neither, preferring to spend their time reading novels instead.[39] Nineteenth-century novelists seem to have made something of a running joke, no doubt shared and appreciated by their readers, of fiction's evil reputation for corrupting young women and encouraging neglect of the womanly arts.[40]

Mending and darning could be temporarily postponed by stuffing the items being repaired into cupboards when callers appeared, but not entirely eliminated, even in a middle-class household.[41] In her 1836 *Young Lady's Friend*, Eliza Farrar (1791–1870) offers advice to young women on the attitude adjustment required for shouldering their share of the world's mending burden:

Most girls consider it a settled thing, that darning stockings is the worst of drudg-ery and, without entering at all into the merits of the case, they cultivate an unrea-sonable dislike to it. This prejudice is often handed down from mother to daughter; and, as it is a business which quickly accumulates on being neglected, the basket of unmended stockings is the dread of all the household . . . ladies who do no other plain work, mend their own stockings. If you look with contempt upon this branch of female industry, and darn your stockings in a great hurry, just when you want to put them on, it will always be an irksome task; but take a pair of stockings in hand, when nothing presses you for time, and darn them whilst listening to read-ing or conversation, and you will find it one of the most agreeable of mechanical employments.[42]

Farrar assumes that it will be a father or brother doing the reading aloud in the family circle on a winter evening; men are referred to throughout her book, with refined ladylike irony, as "the lords of the creation."[43]

Darning is one of the forms of low-status housework undertaken by Dorothy Canfield's role-reversing character Lester Knapp, in her strange 1924 novel *The Home-Maker.* The hostile reception in the community of Lester's home-making role, including his darning of socks, inspires him to opine, at the close of the novel, that "the fanatic feminists were right after all. Under its greasy camouflage of chivalry, society is really based on a contempt for women's work in the home."[44]

Darning, then, was traditionally been regarded in the light of a chore, closer in status to laundry and washing dishes than to embroidery and lacemaking.[45] Hotel maids were expected to darn and mend for guests, as were boardinghouse keep-ers.[46] There are, of course, some documented counterexamples to the stereotype of needleworkers who hate darning and mending, some of them suspiciously fictional, including the characters of Lidia in Mary Doria Russell's 2005 novel of Italy in World War II, *A Thread of Grace,* and that of Four Souls in Louise Erdrich's novel of the same name.[47] Most, however, would have sympathized with Mary Walker's 1846 lament: "Mending, mending, day after day, stitch, stitch."[48]

During the Depression, economic pressures brought about a brief revival of darning, including instruction manuals for those whose former prosperity ren-dered them unfamiliar with the art.[49] During World War II, in all combatant countries darning was regarded as a patriotic duty, however onerous; in the United States the slogan was "Use it up, wear it out, make it do, or do without." Clearly, every mending stitch in a used garment was a blow struck for victory, a comforting and reenergizing context for what would have been a chore in peacetime.[50] For some, like Ursula Hegi's character in the 1994 novel *Stones from the River,* mending is something constructive and distracting for the needleworker to do while her world falls apart around her. As Emily Young's character Mrs. Fraser says in Young's 1947 novel of World War II, *Chatterton Square,* while she kneels to pin a skirt that needs hemming, "I can't make a war or stop one, but I can make you look less of a fright."[51]

Soldiers must themselves have often fought this sense of helplessness with the industry of personal maintenance. The small pocket sewing and mending kits, usually homemade, carried by soldiers in the American Revolution, the War of 1812, the Civil War, and every British war since steel needles began to be made in Europe, was called a "housewife" or "huswif," as if it were a substitute for the woman who would have reattached her man's buttons at home.[52] The luxury of a man's ignorance of darning when among his home comforts is vividly evoked by the British novelist Betty Miller. In her *On the Side of the Angels,* set during World War II, the chief protagonist's Guardsman husband, Colin, is grumbling about his day, drinking cooking sherry *faute de mieux* in the living room, while Honor watches him over her mending basket: "She looked around for her needle, her wool. Slipping the wooden mushroom into the heel of a sock, she began to darn; pricking the long needle in and out: resuming that essential maintenance and repair work, emotional no less than practical, which derives from the feminine desire to preserve at all costs, the *status quo.* Colin, familiar with the sight of her thus engaged, and through familiarity no more conscious of her than of the reconstructive work of his own tissues, walked restlessly up and down the room."[53]

Knitting, like plain sewing, mending, and darning, was a late bloomer in the hedonizing landscape of the needle arts in the eighteenth and nineteenth centuries. Established as an inexpensive method of making footwear and other garments in the Middle Ages in Europe, it remained mostly in the hands of professionals and amateur knitters-of-necessity until late in the nineteenth century. The exception, as we have noted, was the knitting of silk mitts and gloves, purses, and lace in very fine gauge by prosperous ladies like Elizabeth Drinker, of which we have evidence from the late eighteenth century forward.[54]

The origins of knitting are controversial, as it is closely akin to the single-needle process called "nalbinding," and fabrics produced by these means are difficult to distinguish.[55] Rudolf Hommel attributes the origins of knitting to China and argues that Moors introduced the technique to Europe by way of Spain.[56] Knitting on two needles may be an Islamic introduction. Some of the oldest fabrics thought to have been knitted this way, at the excruciatingly fine gauge of 36 stitches to the inch, were found in the Middle East, dating from the seventh to the ninth centuries AD.[57]

European knitting appears to be a relatively late development, with proposed dates of introduction and diffusion varying widely from the early to the late Middle Ages.[58] Considered a plain work technique until the late eighteenth century, it had less demanding requirements for light and good eyesight (or good eyeglasses) than did either embroidery or lace; indeed, the author's blind great-grandmother knitted, although she did not attempt elaborate patterns or, of course, color changes.[59] Some knitters can even knit and read at the same time.[60]

The point of divergence between knitting-as-work and modern hobby knitting had its origins in the sixteenth century, although it was almost three hundred more years before the transition was complete. Crochet and tatting, the former a closely related art, were, as we have seen, leisure handcrafts—fancywork—almost exclusively, from the time of their appearance in Western culture. The overwhelming popularity of modern knitting compared with crochet has tended to obscure this protoindustrial divergence, first of knitting, a method of vertical interlooping, from knit-goods manufacture, and later from related techniques like knotted looping (tatting), and crochet, formed from vertical and lateral interlooping. I rely here, as do most modern textile scholars, on Irene Emery's 1966 classification of fabric structures.[61]

The long slope of hand knitting's diverging path from manufacture resembles that of home sewing much more closely than that of any other needlework technique. Home sewing, however, was even more resistant to hedonization than was knitting, as we shall see in the next section, and persisted much longer as a chore; and home seamstresses embraced the sewing machine nearly as quickly as clothing manufacturers did, while home knitting machines have never been popular, although they have been available for over a century.[62] I do not propose to fully explain why knitting's history differs in so many respects from that of its sister arts; many hobbies and leisure activities have anomalous and conceptually puzzling evolutionary paths, as we shall see in the next two chapters. A few factors, I suggest, may have influenced the long transition of hand knitting from a poorly compensated protoindustrial occupation to a modern path of spiritual enlightenment for the affluent.[63]

Historian of clothing manufacture Claudia Kidwell observes that the design of apparel has much in common with naval architecture: complex and reverse curves present challenges to manipulating a material based on the woven grid in such a way as to allow movement and comfort.[64] Unlike the materials of naval architecture, however, fabrics are supple, and even lightly woven textiles will stretch on the bias (the diagonal relative to warp and weft). Woven fabrics, however, are much less stretchy than knitted textiles, and it is this three-dimensional flexibility of knitting that has made it the structure of choice for stockings, gloves, mittens, caps, and sweaters.

Bias-cut woven fabrics were used as socks and mittens in Europe until the Middle Ages, but once their knitted counterparts were available at affordable prices, consumers from Queen Elizabeth to street peddlers abandoned woven hosiery in droves. By 1774, the knitted product of the English town of Norwich alone was valued at £60,000 annually; and according to economic historian Carole Shammas, nearly enough pairs of stockings were being shipped to the American colonies every year to "put a pair on every adult male's feet," all of them hand- or framework knitted by poorly paid professionals in Scotland or elsewhere in the British Isles. These professional knitters were educated in their art almost exclusively in an oral tradition, supplemented with knitted "exemplars" rather than printed patterns.[65] That handwritten patterns for pleasure knitting, and possibly a few printed ones, were in circulation by 1857 is evident from Charlotte Yonge's novel *Dynevor Terrace*, in which a charming young male dilettante is asked why his "checked shooting-jacket pockets were so puffed out."

> "Provisions for the House Beautiful [the residence of two single sisters]," he said, as forth came on the one side a long rough brown yarn. "I saw it in a shop in London," he said, "and thought the Faithfull sisters would like to be reminded of their West Indian feasts." And, "to make the balance true," he had in the other pocket a lambswool shawl of gorgeous dyes, with wools to make the like, and the receipt [pattern], in what he called "female algebra," the long knitting-pins under his arm like a riding whip.[66]

The urban source of these "provisions" suggests that patterns for leisure crafts were more easily obtained in cities than in rural areas.

The shawl Yonge describes would not, of course, have required individual fitting the way socks and gloves would have. The feet and hands are particularly difficult to accommodate with clothing, for three reasons. First, these extremities are notoriously sensitive to cold and to cold-weather injuries such as frostbite and trenchfoot.[67] Warm, hygroscopic, well-fitting coverings in cold weather are essential.

Second, half of the bones in the human body are in the hands and feet. Anyone who has tried to get a good fit in a pair of jeans, or worse, tried to set a sleeve into a garment, knows that joints and anatomical bifurcations create challenges of fit and comfort, especially if the fabric involved is thick or stiff. The engineering of knitted gloves, for example, must adapt to the five finger separations as well as to the size differences between the wrist and the palm, and the palm to the finger ends. Gussets are often required to permit mobility of the thumb and fingers. Moreover, each person's hands are different in length, shape, width, and flexibility.[68]

Third, because hands are used for work more often than are, say, elbows, covering fabrics need the ability to flex comfortably in every direction. For some tasks, such as archery and shooting, touch is necessary at the finger ends, but the length of the fingers must be covered for warmth, so military mitts are designed with openings at the ends of the fingers. Because the feet of walkers, and, perhaps more importantly, marchers, are in constant and punishing contact with a potentially hostile environment—rocks, snow, rain, sand—their footwear must be free of wrinkles and folds that can produce skin lesions and blisters: these garments must literally "fit like a glove." Socks must fit closely but not be so tight as to restrict circulation, and must be soft enough for comfort yet sturdy enough at the heel turning to resist abrasion from the shoe. Although constructing a garment with seams, knitted flat and then sewn together, can be suitable for sweaters, for example, the hard wear of long marching or walking makes seams anathema in military and other types of heavy-service socks. No other textile structure is as well adapted to these requirements as knitting. Other needlework techniques may be used for decorative effect, but the functionality of garments made from them is limited; one sees crocheted mitts for nineteenth-century ladies of fashion, for example, but not for military sharpshooters. Crocheted socks are a novelty item for modern hobby needleworkers, but again, soldiers do not march in them, nor do skiers and hikers esteem them.[69]

The mathematical complexity involved in engineering knitted garments like socks and gloves, which Kidwell suggests about clothing generally, may have been one of the factors that kept knitting in the category of work long after its sister arts had passed into the hands of hobbyists. Alice Hyneman Rhine noted in 1891 that home knitting of hosiery had been traditional for at least some American women since the eighteenth century and that some still made these commodities at home, although, of course, their numbers were dwindling.[70]

Because most of the European settlement of America took place after the invention of the stocking-knitting machine, the United States has never had a large cadre of professional hand knitters of the type familiar to scholars of the Middle Ages and Renaissance in Europe and Britain.[71] These protoindustrial workers carried their work with them nearly everywhere they went, the Europeans knitting in the characteristically Continental style with the yarn carried on the left (or nondominant) hand and one needle secured in a sheath hung from a belt.[72] The historical evidence suggests that American knitters, never under the production pressure of their European counterparts, did not use sheaths, and as most serious knitters know, knit in a stubbornly inefficient procedure that involves the extra motion of throwing the yarn over the needle from the right (or dominant) hand.[73] Jo March, in Louisa May Alcott's novel of the Civil War home front, may be referring to the more efficient "Continental" method of knitting when she offers to teach the family's new friend Laurie "to knit as the Scotchmen do" to help supply Northern soldiers with socks.[74]

British knitter Mary Thomas, who documented professional knitting in the British Isles as late as 1880, noted that Shetland knitters, using a sheath, could form 200 stitches a minute, twice as fast as the average for an experienced modern hobby knitter.[75] In hedonized production, of course, speed and efficiency are unnecessary. A character in Harriet Beecher Stowe's short story "The Minister's Wooing" (1859) remarks with disapproval on the hedonizing trend of knitting in the nineteenth century: "'Girls a'n't what they used to be in my day,' sententiously remarked an elderly lady. 'I remember my mother told me when she was thirteen she could knit a long cotton stocking in a day.'"[76] This kind of diligence would be entirely superogatory for a hobbyist, though all-day knitting is not unusual for leisure artisans in love with the intended recipient of the garment in question, mesmerized by the mysteries of pregnancy, or simply hypnotized by the rhythm of the knitting itself.[77]

Two developments opened the door to leisure knitting with wool, silk, and cotton in the nineteenth century: the fall in the price of knit goods, especially stockings, as the machine-made hosiery industry expanded, and a parallel fall in the prices of first wool and then cotton yarn and thread for hand-knitting. Quality improved as well, with fine plied worsted yarns available to hobby artisans by the mid-nineteenth century. The invention of aniline dyes increased the variety of colors available, and reduced the price of colors like red and purple, bringing them within reach of middle-class needleworkers by the 1880s.

The knitting needles used by leisure artisans of the eighteenth and early nineteenth centuries were of much smaller diameter than those we see today. Many of them were not much more than wires cut in points at both ends.[78] Sizes for knitting needles and crochet hooks were standardized in Britain and the United States. in the late nineteenth century, although, unfortunately, not to the same standard.[79] Nearly all preindustrial and protoindustrial knitting needles were double-pointed, and point protectors linked by string or chain were necessary to keep stitches on the needles.[80]

Bobbin lace, like weaving, remained in professional hands from its origins in the fifteenth century until the mid-twentieth century, partly, one supposes, because it is a very difficult art to learn, and partly because its portability is limited.[81] Other factors may have contributed to its failure to attract hobby artisans before the 1960s, among them the cumbersome tools and equipment, including hundreds of small, very sharp pins and bobbins, some with glass or bead spangles, both of which must have been nightmares for lacemakers with small children. Bright light is required, and because of the mathematical complexity associated with crossing dozens of independently suspended bobbins over and under each other in a pattern, bobbin lacemaking is often incompatible with casual conversation, supervision of children, or cooking. Scrupulous care must be taken to keep the hands clean and dry because soil worked into the lace will not wash out, resulting in what was, in the world of commercial lacemaking, known euphemistically as "colored" lace. Because professional lacemakers were almost as impoverished as hand-knitters, they were regarded in Europe, and later in the United States, as quaint and colorful human features of protoindustrial economic landscapes.[82]

Professional lacemaking is still carried on in some parts of the world, mainly Asia, but machine-made laces dominate the commercial market.[83] Bobbin lacemaking became visible as a hobby activity in the United States in the 1960s, postdating even revivalist handweaving, which had its origins in the 1920s and 1930s.[84] Earlier efforts to create hobby interest in lacemaking, notably by Louisa and Rosa Tebbs in Britain in the first quarter of the twentieth century and by Margaret Maidment on both sides of the Atlantic in the early 1930s, had failed to sustain more than a brief initial enthusiasm.[85] In the late 1960s however, a small group of Americans, including Osma Gallinger Tod and the husband-wife team Kaethe and Jules Kliot, succeeded in organizing a critical mass of American hobby lacemakers.[86] The Eliots, Europeans transplanted to Pawley's Island, North Carolina, where they ran a hammock and net business, applied modern design principles to traditional lacemaking and macrame techniques.[87] By the 1970s, North American lacemaking had its own international hobby organization, charmingly (and controversially) denominated the International Old Lacers.[88]

Diagnostic Artifacts: The Sewing Machine and the Quilt

The freedom from the endless demands of plain work and mending enjoyed by the needleworker after the mid-nineteenth century was, as we have seen, the product of many factors relating to production and distribution of consumer goods. In this context, it is almost impossible to overestimate the influence of the sewing machine, which revolutionized both the needle trades and home sewing. The sewing machine's liberation of stitchers from the endless sewing of seams, and eventually from the necessity of sewing any seams at all, opened the door to needlework's possibilities as a recreational and artistic medium.

In the eighteenth century and the first two decades of the nineteenth, most clothing in the United States and Britain was still made at home, some of it by tailors and dressmakers who were hired into the household temporarily to complete or augment the family's wardrobe.[89] For major life events requiring new clothes, especially those involving time constraints, such as weddings and funerals, professionals were nearly always called in to help complete the new wardrobe. Merchant tailors, dressmakers, and milliners made clothing to order from hand-drafted patterns, cutting, fitting, and sewing one garment at a time for those who could afford their services.[90] In the United States, tailors cut and fitted garments to customers and then, in many cases, sent packages of pieces with thread and buttons to home seamstresses in surrounding areas.[91] Since women were paid piece rates much lower than those of men, this practice reduced labor costs considerably.[92]

During the 1820s, some merchant tailors began to make use of slack time in their shops to produce cheap, unfitted garments for sailors, workers, slaves, and other persons whose station in life did not permit pretensions to *haute couture*.[93] This proved so profitable that production of ready-to-wear, including the so-called "Southern work," became the specialty of some men's apparel houses. By the time the sewing machine became a marketable invention in the 1850s, significant demand already existed for it in the thriving garment industry, which sought to reduce labor costs and to profit from the rapid market expansion in ready-to-wear.[94]

Like many other technological innovations, the mechanization of sewing was highly controversial in its time. Some observers in the mid-nineteenth century expected the sewing machine to create massive unemployment in the needle trades.[95] Since the labor requirement in hours per garment was reduced by more than 90 percent, these critics argued, the jobs of nine of every ten seamstresses and tailors would be rendered obsolete. Market expansion between 1850 and 1880, however, refuted these pessimistic predictions, as employment in garment manufacture more than doubled in thirty years.[96]

The farm-based American population of the eighteenth and early nineteenth centuries had required sturdy, comfortable costume for daily wear and perhaps one set of dressier garments for church, weddings, funerals, and the like. The dressier clothes were often made with the assistance of professionals. After the mid-nineteenth century, however, urban areas had new kinds of work and new workers, whose family structures did not always permit continuation of home production traditions. Women in the labor force, for example, did not always have the time or space to make clothing for themselves and their families. In addition, many men took employment which very nearly precluded family life entirely, such as in the maritime industry, whose workers were the first major market for ready-to-wear, and the mines, logging communities, and ranches of the western United States. For much of this kind of work, the most practical clothing was heavy, very difficult to sew by hand, and, most importantly for the garment industry, loose-fitting by comparison with tailored fashions. Loose fit made it possible to manufacture such garments without close fitting to individual buyers, and the construction of heavy, rugged work clothing was an ideal task for the sewing machine.[97]

Another growing market in the mid-nineteenth century was the urban middle class of white-collar workers, whose status was aptly described by reference to their dress. Those who worked in the upwardly mobile ranks of businesses, banks, and government bureaucracies were expected to dress like gentlemen and gentlewomen on the salaries of clerks, appearing daily in fresh shirts (or, at least, fresh collars and cuffs) and respectable suits.[98]

Early in the nineteenth century, in part due to the astonishingly broad influence of the British fashion leader George Bryan ("Beau") Brummel (1778–1840), suits for both sexes, worn with clean, starched shirts and frequently changed underwear, had come into fashion both in Britain and in the United States. Initially, as is nearly always the case with fashion, manifested by the nobility (in Britain) and the wealthy (in the United States), the demand for frequent changes of both inner and outer clothing spread quickly to the urban middle class on both sides of the Atlantic—a godsend to the garment and laundry trades.[99] This sartorial democratization would have been impossible without the introduction of low-priced, mass-produced clothing. Apparel sizes for men were introduced in the United States during the Civil War, and various schemes of size grading and garment cutting developed in the second half of the nineteenth century. The first home sewing pattern, for a shirt, was published by Butterick about 1850.[100]

While the sewing machine was marketed to home consumers as early as 1856, it was designed originally as a manufacturing tool for the garment and shoemaking trades.[101] The sewing machine drove handwork out of the industry in all but a very few limited areas of production, such as tailored buttonholes. Millinery, felt and straw hat making, and artificial flower production were slower to mechanize than garment manufacture, but these trades, too, soon incorporated the sewing machine into many tasks previous done by hand.[102]

The burden that the sewing machine and industrialization of garment production lifted from women who had previously sewn their families' clothing by hand may be judged by an 1869 article in *Harper's Bazaar*:

> The following curious calculation of the number of stitches in a shirt, which somebody has had the patience to make, we mean the calculation not the shirt, by any means, may induce some gentleman to present his wife with a good sewing machine. Stitching the collar, four rows, 3,000; sewing the ends, 500; buttonholes and sewing on buttons, 150; sewing the collar and gather the neck, 1204; stitching wristbands, 1228; sewing the ends, 68; buttonholes, 48; hemming the slits, 264; gathering the sleeves, 840; setting on wristbands, 1468; stitching on the shoulder straps, three rows each, 1880; hemming the bosom, 393; gussets, 3050; tapping the sleeves, 1526; sewing the seams, 848; setting side gussets in, 424; hemming the bottom, 1104. Total number of stitches, 20,530.[103]

As a domestic appliance, the sewing machine was quite expensive, costing as much as $125, an obstacle that Singer addressed by a financial innovation in 1856, the installment plan, later to be applied to almost every type of consumer durable.[104] Sewing machine historian Brian Jewell, stretching a point, asserts that "apart from the clock, the sewing machine was the first industrially produced domestic mechanism." So it is hardly surprising that new methods of retailing and financing had to be devised before the sewing machine could be successfully marketed to consumers. The sewing machine created another difficulty that required a technological resolution: the hand-sewing thread available in the 1850s was not smooth enough to pass through the sewing machine's tension mechanism and needle. It was not until the mid-1860s that the introduction of Clark's ONT ("Our New Thread"), a hard-twisted six-ply product, solved the problems of breaking and snagging.[105]

Historian and curator Grace Rogers Cooper says that "the sewing machine became the first widely advertised consumer appliance, pioneered installment buying and patent pooling, and revolutionized the ready-to-wear clothing industry."[106] It did more, however. Although the sewing machine was sold both to consumers and to industrial customers in the second half of the nineteenth century, its advent separated the two permanently and made of them two distinct classes of needleworkers.[107]

For the consumer, the sewing machine and the existence of the garment industry made home sewing much easier and, more importantly, made nearly all of it entirely unnecessary. The home needlewoman could pick and choose which sewing and needlework tasks she enjoyed performing and leave the rest to her sisters in the apparel industry. Even machine embroidery has enjoyed only limited popularity among hobby needleworkers, despite strenuous efforts by sewing machine manufacturers to encourage it.[108] One of the sewing tasks the leisure needleworker evidently did want to continue performing was the making of quilts, to which we now turn.

The hedonization of production technologies occurs along an advancing boundary of material abundance created by the very success of these technologies: because we do not need to make essential objects ourselves, we are free to make them for pleasure if we feel like it. Nowhere is this more evident than in the history of patchwork, appliqué, and quilting. Eliza Calvert Hall's character Aunt Jane described in 1907 both the transition from necessity in home sewing and the hedonizing character of quiltmaking in the late nineteenth century:

> There never was any time wasted on my quilts, child. I can look at every one of 'em with a clear conscience. I did my work faithful; and then, when I might 'a' set and held my hands, I'd make a block or two o' patchwork, and before long I'd have enough to put together in a quilt. I went to piecin' as soon as I was old enough to hold a needle, and a piece o' cloth, and one o' the first things I can remember was settin' on the back door-step sewin' my quilt pieces, and mother praisin' my stitches. Nowadays folks don't have to sew unless they want to, but when I was a child there warn't any sewin'-machines, and it was about as needful for folks to know how to sew as it was for 'em to know how to eat, and every child that was well-raised could hem and run and backstitch and gether and overhand by the time she was nine years old.[109]

As Hall's character implies, quilting combines what are, in fact, three separate subtechnologies of hand sewing that converged in the late seventeenth and early eighteenth centuries in Europe and a little later in the American colonies: the first, the stitching together of pieces of fabric with ordinary seam stitches, such as back stitch or running stitch; the second, a method of laying one piece of fabric on another and tacking it down with any of a number of possible stitches; and the last, an ancient technique for making two pieces of fabric more than twice as warm as one by sandwiching an insulating material (wool, cotton, feathers, or down) between them and securing the whole with closely aligned running, double-running, or back stitches. The Chinese discovered that not only comfort but life itself could be preserved inside a quilted garment and were using it for armor before Europeans had steel needles.

In the late eighteenth and early nineteenth centuries, several lines of economic and cultural influence were converging to produce a characteristically American product: the patchwork quilt. While other nations and ethnicities have produced quilted products of great interest and beauty, the American quilt is a diagnostic artifact in the anthropological sense, like the costume embroidery of late-nineteenth-century Bukovina and the Irish-American crochet lace of the 1920s.[110] It blossomed in the margins of plenty, while yet serving a purpose that exculpated its maker from idleness, as "Aunt Jane" observes.

Among the quilting and patchwork traditions European-Americans brought to the New World were those of Britain, where quilted fabrics had been made as a type of plain work since the Middle Ages.[111] Much of the quiltmaking in America began as plain work as well, but by the mid-nineteenth century, the craft was already undergoing hedonization.

The evolution of the quilt from the eighteenth century through the twentieth was shaped by technological and economic developments that gradually altered the design paradigm for artisanship from one of scarcity of materials to one of abundance, a typical pattern of change for the textile arts. The mechanization of production processes from cotton ginning to sewing permitted needleworkers to decide for themselves how much of the work of quiltmaking they wanted to do themselves.[112]

Before industrialization, cloth and clothing were scarce and so expensive that few could afford to own much of them. Many clothing designs were "peasant-cut" on straight grain lines to economize on fabric, leaving few cutaways to incorporate into quiltmaking.[113] Preindustrial quilts were often made of whole or uncut cloth, but sometimes included recycled components of petticoats or furnishings.[114]

Printing technology, too, was a strong influence on quiltmaking. The printed textile called a palampore, a large Indian coverlet or hanging popular in the seventeenth and eighteenth centuries, inspired numerous less-expensive imitations in textiles embroidered and/or quilted at home.[115] The technique called *broderie perse*, which employed appliqué motifs cut from printed textiles to embellish quilt tops and other textiles, was a similar effort to reproduce the effect of an expensive printed textile by arranging smaller pieces in a larger design that emulated the luxurious effect.[116] Cotton became an inexpensive material by the middle of the nineteenth century, and with the invention first of aniline dyes in 1857 and later of innovations in screen and roller printing of fabrics, printed cottons became an affordable medium for even economically disadvantaged American quiltmakers.[117]

The Industrial Revolution not only reduced the prices of materials relative to wages but also made tools and patterns more affordable. As we have seen, the prices of needles, scissors, thimbles, thread, lighting, and even comfortable furniture for sewing fell steeply through the nineteenth century.[118] Hard-twisted cotton sewing thread, invented in the early nineteenth century and reengineered in mid-century for use in sewing machines, drove its linen predecessor out of the home sewing market by the end of the nineteenth century.[119]

All of these developments combined to enhance the pleasure of making quilts. Needleworkers had the option of stitching the plain seams on a machine, and even for hand stitchers the smooth, even cotton thread available after 1850 slid through fabrics with far less effort than the slubby homespun linen their grandmothers had used. British historian of needlework Therle Hughes says of the joys of piecing that "patchwork is unique in that it offered the hardworking cottage woman as well as the over-leisured painstaking girl in her teens an opportunity to revel in colours and textures of fabric far removed from their workday world—and how they revelled."[120] The American antiquarian Alice Morse Earl had used a similar rhetoric of pleasure in her 1898 *Home Life in Colonial Days*: "The feminine love of color, the longing for decoration, as well as pride in skill of needlecraft, found riotous expression in quilt-making. Women reveled in intricate and difficult patchwork."[121] Even the Old Order Amish reveled, despite their religious restriction to plain (unprinted) colored cloth; there was no prescribed limit to the number or brightness of the solid colors these artisans could use, and Amish quilts are among the most visually striking of all American quilts.[122]

The printed patterns for sewing, piecing and quilting that were available after the 1850s made it unnecessary for needleworkers to draft their own or to purchase designs from "stampers," whose prices had reflected the skills and labor-intensivity necessary for pattern drafting.[123] There was still a great deal of hand work in quilts, however, at every phase.

The cutting of quilt pieces from fabric is a precision operation; mistakes produce warped, bunched, or wrinkled patches that will not lie flat when set together. Each piece must have a seam margin, usually of fixed width, which must be followed during the piecing and setting together of patches.[124] Appliqué does not have such close tolerances, as pieces are laid on top of each other rather than fitted, but careful measurement and cutting are still required. Templates, the patterns quiltmakers use to cut pieces, historically of any and every relatively stiff material from writing paper to sheet metal, typically do not show these allowances, so quilters usually must cut them by eye. Because quilting involves sewing many very small patches together, with considerable risk of not "tiling the plane" evenly, needleworkers of the nineteenth and early twentieth centuries ensured the flatness and uniformity of their pieces—sometimes tens of thousands of them—by lining each piece with paper, a less stretchy medium than fabric, felling down each tiny seam around the individual paper linings of the patch. The paper was sometimes left in, a practice that was a nightmare for laundry purposes and a godsend to historians, who can sometimes date quilts by the paper incorporated into them.[125]

After completion of the top, the quilting pattern had to be marked onto it, either with quilting templates for fancy designs or with a chalked line "snapped" on at measured intervals, after which the three components of the fabric sandwich were carefully basted together by hand.[126] Although it is, and was after 1850, possible to quilt by machine, this method reduces the "loft" of the quilt—its thickness and dimensional quality—and so hand quilting persists among modern artisans, more than a century and half after the invention of the sewing machine.

American patchwork quilters have always had a certain ambivalence about the nonportability of the end phase of quiltmaking, when the fabric sandwich must be stretched on a frame large enough to accommodate its entire width, although the length may be reduced by the use of a roller frame. Stretching the arms across the quilt could become quite painful after several hours of quilting, and by the time the quilter "shook hands" with her counterpart on the other side of the frame, hands and arms might be aching or numb.[127] Patchwork, or "piecing," however, is and was readily portable, can be picked up and dropped at will, and could even be carried around in the needlework aprons, called "workpockets," popular in the eighteenth and early nineteenth centuries. These devices, often patchwork or embroidered themselves, allowed the needleworker to carry her tools with her without encumbering her hands and, perhaps more importantly, to stuff the work-in-progress quickly into the opening in the front of the bag when a child strayed too close to the fire or dinner threatened to stick to the pot. Piecing of bags, work pockets, doll clothing and bedding, pincushions, and other small articles were popular ways to use up clothing cutaways and to make creative use of leisure time. In Charlotte Yonge's 1861 novel *Hopes and Fears*, one of the characters makes use of the enforced leisure of illness by piecing 139 patchwork bags, each destined for a particular recipient.[128]

Carrie Hall, in her foreword to the 1935 *Romance of the Patchwork Quilt in America*, describes becoming so engrossed in the art of quiltmaking that, after making thirteen quilts, she realized that it was simply not feasible to make a full-sized quilt of every known patchwork pattern, however appealing the idea might be: "When the World War [I] was over and there was no longer a necessity for knitting socks and sweaters, I found my fingers itching for some 'pick-up' work, so I turned to quiltmaking . . . After completing my 'baker's dozen' I realized that I couldn't continue making quilts indefinitely, and yet I was so fascinated by all the numerous and beautiful patterns that I conceived the idea of making a collection of patches, one like every known pattern, little realizing the magnitude of the undertaking. The collection now contains over one thousand patches and is to be placed in the Thayer Museum of Art of the University of Kansas."[129] This corpus of patches remains at the University of Kansas, a "labor of obsession" that has now been appreciated by several generations of quiltmakers and textile historians. [130]

Another type of "pick-up work" that utilized scraps and remnants was the making of doll clothes, usually, although not always, done for pleasure. While some schools used the stitching of tiny garments for the limbs of dolls as a cheap, portable, and effective educational technique for plain work, most doll clothes made at home were sewn for pleasure. Helen Campbell's 1883 *American Girl's Home Book of Work and Play,* for example, devotes fourteen pages to the delights of "doll's dressmaking" but only a quarter-page apology for so much as mentioning "plain sewing."[131]

The End of Necessity

Needlework was only one of many branches of artisanship that bifurcated into hobby and production paths in the eighteenth and nineteenth centuries, as crafts were first mechanized and then industrialized. While historians and sociologists of labor are inclined to romanticize artisan production, including home manufacture of necessities like soap, candles, and socks, there is ample evidence that the practitioners of these arts often did not experience the rosy glamour these tasks have taken on in hindsight.[132] Some women hated needlework, some men hated carpentry, and many crofters hated gardening, just as, in the period before automobiles, many persons hated horses and all their works, although they had no option but to make use of them anyway. Industrialization and economic democratization allowed those who disliked crafts or other activities to drop most of them entirely, and freed those who did like them to explore their potential for pleasure. Moreover, the technological change associated with industrialization introduced new kinds of technology-dependent hobbies, such as photography, model railroading, and ham radio, and encouraged the kind of popular enjoyment of amateur science and engineering that had been associated only with wealthy elites in the seventeenth and eighteenth centuries.

While, as we have seen, hobbies and leisure activities significantly predated the nineteenth century, this was certainly the initial period of their rapid and widespread democratization. By 1890, there was a prosperous and expanding market in leisure needlework patterns, tools, and materials, and nascent hobby markets for many other crafts as well. Home artisans in metal and wood were discovering the pleasures of owning and operating a lathe; fretsaws were popular as hobby tools for both sexes; and amateur photography amounted to a veritable craze that has persisted to this day. How these markets developed and adjusted to increasingly leisured consumers is the subject of the next chapter.

CHAPTER FOUR

The Hedonizing Marketplace

So many technologies became hedonized in the United States and Britain between the middle of the nineteenth century and the end of the twentieth that it would be nearly impossible to enumerate them all, or to do justice to the explosion of ideas, not only for the uses of leisure time on the part of consumers, but for developing markets of tools, materials, and publications on the part of authors, publishers, designers, and manufacturers. While I intend to retain my focus on needlework as the case with which I am most familiar, my objective here is to survey briefly the flourishing marketscape of leisure crafts and hobbies between the middle of the nineteenth century and the end of World War II, and to situate needlework within this context. I hope to see woodworking, ceramics, photography, and the many other hedonizing technologies that are addressed in the latter part of this chapter more fully explored by those who understand them better; my objective here is merely to point in the direction of rich troves of primary sources available to the scholarship of leisure activities and crafts.

I have left the arena of competitive sports as a use of leisure to the historians of this well-documented human activity, although a few of the hobbies I have touched on in earlier chapters, such as hunting and fishing, are usually classified as sports. I have included them because they are male analogs to leisure textile arts, and I will discuss them and related activities briefly here in the same context. As in earlier chapters, I will draw on bibliometrics to sketch out the shape of developing markets in leisure crafts, in part because comparable data are available for a large number of hedonizing technologies, and partly because I have never been able to resist the lure of popular publications as a source of such entirely implausible facts as the presence of electromechanical vibrator advertising in the pages of the needlework magazine *Modern Priscilla* in 1910. The texture of the past is always much richer than we historians can capture, and popular literature is one of many ways to catch glimpses of this incommensurable complexity.

As we have seen, artisan technologies enter the leisure market at different times and hedonize at different rates, even within the same general craft area, such as needlework. We have seen, for example, that embroidery and sewing, which appear similar to uninvolved observers, have taken very different paths into leisure activity and are centuries apart in the timing of this transition. They acquire what I have called here the second and third derivatives at different times as well: the demand by artisans for information (including, often, historical information) about their craft, which stimulates and supports publications, followed, or sometimes accompanied by, a demand from collectors, connoisseurs, and museums for examples of and publications about what they consider to be the finer elements of the craft.

The Market for Tools and Materials

As industrially produced craft tools and materials began to be economically accessible to a broader range of social strata in the nineteenth century, the focus of marketing efforts for them shifted to emphasize the hobby artisan's enjoyment of their use. Needlework techniques such as floral embroidery, once the province of professionals on the one hand and the leisured and prosperous on the other, became a mainstream leisure activity in middle-class and upper-working-class markets. Crochet and tatting were thriving at nearly all economic levels in the United States and the United Kingdom by the end of the nineteenth century, and knitting was well along in the process of hedonization at the upper end of the social scale with silk and lace knitting.

Significant to all these developments was the fall in prices of tools and materials. Cotton sewing thread, both hard- and soft-twisted cotton yarn, and thread products for crochet, knitting and tatting were new industries in American cities like Pawtucket, Rhode Island, and Willimantic, Connecticut. Silk and cotton yarns were spun in Belding, Michigan; New London, Connecticut; and Philadelphia, which was also a center of wool knitting yarn production. Although linen had not disappeared from the needlework market, its demographic had changed, and as a plainwork sewing thread it lost market share among hobbyists as cotton thread fell in price through the nineteenth century, protected in the United States by a stiff tariff on the imported product. The thread and notions industry was thriving from the 1870s through the mid-1920s, when the textile depression drove many companies out of business and forced mergers of others.[1]

Needlework tools, other than sewing needles, which had become fully industrial products more than a century earlier, began to be available at attractive prices in variety, department, and dry goods stores at the end of the nineteenth century and the beginning of the twentieth, as well as by mail order. Embroidery frames were turned out on industrial lathes and assembled from manufactured hardware.[2] Companies like C. J. Bates of Chester, Connecticut, established in 1907, turned out crochet hooks and knitting needles from wood and sheep bones on lathes, cutting the hooks into the heads by hand. Milward, a British needle and needlework-tool manufacturer whose products were and are popular with American leisure artisans, made hooks and needles of celluloid, wood, bone, and steel. Some specialized tools were invented, such as the type of spool-holder suspended from a wrist ring favored by crocheters using fine cotton and silk threads.[3]

Tatting shuttles were manufactured from metal, celluloid, bone, or, more rarely, ivory. Knitting needles acquired in the nineteenth century their modern appearance, with caps on one end of the type of needles used for flat knitting. Needles for knitting in the round, as for socks, remained double pointed, and were (and are) usually sold in sets of four; some of the uses for these artifacts were taken over later in the twentieth century by the so-called "circular" needle, essentially a double-pointed needle with two ends connected by a long, thin, flexible shank.

One of the most important developments of the late nineteenth century was the standardization of needle and hook sizes, an innovation that made it much easier to buy appropriate materials and to follow written instructions for needlework.[4] Hand-sewing and embroidery needles, which were made in Britain but not the United States, were standardized much earlier in the nineteenth century than other needlework gear. Thimbles on both sides of the Atlantic were traditionally made to ring sizes, and when they began to be manufactured in large quantities, these sizes were usually stamped on the product.[5]

Inevitably, knitting needle and crochet hook sizes were standardized differently in the United States, Britain, and continental Europe.[6] Needleworkers using a pattern from abroad usually had to make the needle size conversion by knitting or crocheting a gauge sample, a small square or rectangle to determine which size needle or hook would produce stitches in the chosen yarn of the size that would result in a product of the appropriate size. Before 1929, even this procedure often had to proceed by guess and by golly, as not all published patterns provided a gauge standard. The 1847 *Winter Gift for Ladies*, for example, told its readers to knit a baby's muffler "in German Wool, with four needles No. 19. Cast on 53 stitches . . . " Later, the anonymous author suggests "very coarse" needles for another project. The very small "Work Department" in *Godey's Lady's Book*— a section that ran a maximum of six to eight pages per issue—rarely provided any but the most sketchy instructions and even more rarely credited the designers of published patterns. Miss H. Burton's 1875 *Lady's Book of Knitting and Crochet* mentioned needle sizes, but most patterns had no indication of gauge except rough weight categories for yarns, for example, "fine" or "coarse." Marie Louise Kerzman, who authored a knitting manual in 1884, did not even bother to provide information about needle sizes for patterns, except those for rugs, and straightforwardly punted the gauge issue, which she asserted was left "to the judgment of the knitter."[7] Clearly, readers were expected to work up gauge samples before beginning their projects. (Gauge samples are still recommended before beginning work on a new pattern, as individual differences in knitting or crocheting style can significantly alter the size of the finished article; some needleworkers save these gauge samples as stitch exemplars, a practice advocated by Barbara Walker's magisterial series on hand knitting.[8]) Since not all knitting needles were marked with sizes (double-pointed and circular needles still are not), knitters used gauges with sized holes to determine which of their needles were of the proper size for the intended result.[9] Small retailers of needlework materials frequently offered assistance with these difficulties, as they still do today.

Patterns, tools, and materials for needlework were sold to American consumers after 1880 in dry goods stores, department stores, general stores, variety stores, some retail millinery and dressmaking establishments, and by mail order.[10] The six-cord cotton thread that was most widely used for crochet, knitting, and other hobby needlework was available from wholesalers for 40 to 45 cents per dozen in 1892. Cotton sewing thread was sold to consumers on birch spools for five or six cents a spool in 1909.[11]

Virginia Penny, in her 1870 book *How Women Can Make Money*, noted that the market for needlework tools, materials, and patterns was one in which women had an opportunity to make successful careers in retailing, owning, or managing fancy goods stores and variety stores.[12] City directories of the period in the northeastern United States show many women doing just that, sometimes offering services as embroidery stampers and/or designers in addition to selling the products that leisure needleworkers needed for their craft. Many dressmakers and milliners, including the large number of foreign-born in this occupation at the turn of the century, sold needlework materials and sometimes educational and design services as well.[13] In New England, the directories for Lowell and Lawrence, Massachusetts; Manchester, New Hampshire; and the Pawtucket/Central Falls area of Rhode Island show a boom in most of these types of retailing at the turn of the twentieth century, followed by a decline after 1910.[14]

The Market for Publications and Designs

As in the case of hobby woodworking and gardening, an expert literature of needlework for pleasure began to appear in Europe, Britain, and the United States about 1840 and reached boom proportions by 1920. The form of the patterns changed as well; by 1900, the majority of American and European needleworkers were literate and expected text instructions as well as illustrations in their patterns. In the case of some types of needlework, the technical algorithms shifted. Aran knitting, for example, shifted between 1890 and 1940 from a conceptual strategy based on preliterate methods of working in consecutive vertical panels to one of conceptualizing an entire garment piece in terms of rows.[15] As literacy and numeracy became increasingly widespread, patterns could be captured in pictures and text, and instructions spelled out in detail, allowing hobby artisans much greater access to the collective memory of designs and techniques, including those of ethnicities other than their own. Thus, neither textile exemplars nor living artisans providing narrated demonstrations were needed to carry on craft traditions, both developments with parallels in other leisure activities. Standardization of gauges, abbreviations, and stitch terms, still an ongoing international project, created crosswalks between different national and ethnic traditions.[16] Some publishers of turn-of-the-century instruction books emphasized this transition from in-person demonstration by calling their books "self-instructors."[17]

Berlinwork embroidery, or wool embroidery on counted canvas, a popular art in the nineteenth century and much despised as frivolous and ugly by later critics and connoisseurs, was worked from charts, which were themselves proto-industrial artifacts.[18] Designers of this type of embroidery drew the overall design—or sometimes only a quarter of it—onto printed point paper, after which each square of the paper, corresponding to an intersection of warp and weft in the ground canvas, was laboriously painted in by hand in colors corresponding, in very general terms, to those of the embroidery yarns.[19] Curves had to be stepped. The labor-intensivity of these patterns, which were produced in Germany as early as 1804 and remained in the market until late in the nineteenth century, made them expensive. To keep costs down, some were drawn and painted in only one quadrant; the needleworker used either a mirror or her imagination to stitch in the other three-quarters of the pattern. When printed in the needlework and household magazines that had begun to appear by 1830, the charts were represented in black and white, and color choices left to the artisan.[20] Berlinwork

Hand-colored Berlinwork embroidery pattern published in Germany in the mid-19th century, representing only one quadrant of the design. Later critics scorned this type of work despite the obvious skill required to execute the curves and shading, to say nothing of using mirrors or the imagination to reconstruct the other three quadrants.

charts were among the first commercially printed needlework patterns to be distributed internationally in a nonbook format.

Berlinwork was not only a target of criticism from those claiming elevated taste in art; it—and indeed other types of needlework—was also a favorite whipping post for feminists and nineteenth- and early-twentieth-century leftists generally. Mary Braddon (1835–1915), describing her fictional character Gwendoline Pomphrey in 1864, referred to this sociopolitical trope: "Her beauty, a little sharp of outline for a woman, would have well become a young reformer, enthusiastic and untiring in a noble cause. There are these mistakes sometimes—these mésalliances of clay and spirit. A bright ambitious young creature, with the soul of a Pitt, sits at home and works sham roses in Berlin wool; while her booby brother is thrust out into the world to fight the mighty battle."[21] A "trades-unionist" in Margaret Oliphant's 1881 novel *He That Will Not When He May* dismays a group of aristocratic ladies embroidering in a well-furnished parlor by contrasting their situation to that of underpaid seamstresses resorting to prostitution when they cannot get sewing to do: " 'Great god!' cried the orator, jumping up. 'Why should we be sitting here in this luxury, with everything that caprice could want, and waste our lives working impossible flowers upon linen rags, while they are starving, and perishing, and sinning for want, trying for the hardest work, and not getting it?'"[22] Feminist and social critic Lorine Pruette (b. 1896) also disapproved of hedonized needlework, writing in 1924 that she would have preferred to see girls and women exercising in the fresh air, preferably in the nude: "Mary Ann of the twentieth century knits through interminable miles of colored wool and fills her house with crocheted and embroidered atrocities, just as her Polynesian sister flashes her brown body through the warm seas; the one spoils her eyes in the preparation of a beaded bag, while the other gathers hibiscus flowers to deck her dark hair."[23]

Designers and Publishers

Feminist critics of needlework in this period were apparently oblivious to or uninterested in the opportunities that needlework and other crafts were creating for women as designers, and as authors and/or editors of hobby crafts publications. Turn-of-the-century women who benefited from the hedonization of needlework were aware of the trend as an advantage to themselves and their dependents, as articles in crafts magazines amply document, even in articles on plain work for pay.[24]

One of the most prominent pioneers in the international and multicultural enterprise of hobby needlework in the late nineteenth century was the Viennese author and needleworker Thérèse de Dillmont (1846–90) of Dollfus-Meig et Cie. (DMC), an Alsatian-French textile firm established in 1746 in Mulhouse to produce hand-painted fabrics, which is still producing and selling craft yarns and thread at this writing. Dillmont's name appeared on more than a hundred publications in seventeen languages after 1870; DMC continued to issue titles after her death under the imprimatur of her niece, reportedly (though implausibly) also named Thérèse de Dillmont.[25] The elder Dillmont's most famous work, known in English as the *Complete Encyclopedia of Needlework*, is still in print and remains one of the most comprehensive practical treatments of the needle arts. Published by 1900 in French, German, Spanish, and Italian, the *Encyclopedia* effectively launched DMC as a major global publisher of needlework patterns.[26] A Russian edition of the work appeared in 2004, more than 120 years after its initial publication.[27]

Dillmont and DMC brought together in the former's Dornach studio thousands of patterns from all over the world, which were published over the years as charts and drawings, usually with text instructions, bringing to the global leisure market high-quality algorithms for the designs and techniques of French lace,

Irish crochet, soutache-braid laid work, a broad variety of alphabets, fancy knitting, cut and drawn work, the coarse net lace called "guipure," macramé, needle laces, knotted fringes, Teneriffe lace from the Canary Islands, tatting, and embroidery designs from Morocco, Coptic Egypt, China, Turkey, Czechoslovakia, Yugoslavia, Bulgaria, Italy and France.[28] Some of these pattern books included perforated paper design pages for pouncing patterns onto cloth; others, in the twentieth century, had hot-iron embroidery transfers (wax on paper or cloth).

Other publishers and craft supply manufacturers, at the end of the nineteenth century and the beginning of the twentieth, were doing their best to tap into this trend toward hobby-textile multiculturalism as well. Magazines like *Ladies' Home Journal, Godey's, Leslie's,* and *Petersen's* in the United States had been publishing needlework patterns since the third decade of the nineteenth century, and those of Britain and Europe since the late eighteenth century, but these consisted largely of drawings of finished articles, with only the sketchiest of working instructions. Moreover, the illustrations often failed to show stitch structure, a defect that was remedied in DMC's publications and those of other yarn companies by the 1880s through the inclusion of detailed drawings. By the early twentieth century most pattern books included sharp, clear photographic enlargements of stitch details, much prized by needleworkers who may have learned their art from "exemplars" and who now puzzled over the newfangled text instructions, with their inscrutable abbreviations.[29] Pattern publishers to this day are careful to photograph needlework projects, such as doilies, worked from the center out, with the central motif clearly visible.

The hobby market for needlework serials was very nearly global by the 1920s, with publications originating not only in the United States, United Kingdom, and France but also in many other European countries, Japan, and Central America.[30] Embroidery, lacemaking, crochet, and knitting were well represented among the dozens of serial titles that began appearing in the late eighteenth century, but plain sewing was represented only by high-fashion dressmaking and its needlework accessories. Quiltmaking serial titles are entirely absent until well into the third quarter of the twentieth century. Several of these publications directly address the hedonizing market, employing the rhetoric of pleasure in their titles. The London-based *Lady's Magazine*, established in 1770, advertised that it was an "entertaining companion for the fair sex, appropriated solely to their use and amusement." In Brussels, the *Trésor des Demoiselles* began publishing in 1859. If nothing else, the vigor of needlework publishing in the late nineteenth and early twentieth centuries testifies to the amount of buying power publishers and manufacturers saw in the market for fancywork patterns and materials.

The Evolution of Crafts Publishing

In most of the hobby arts, it is possible to track economic change through the demand for publications more readily than that for tools and materials, because the former are more readily distinguishable as to intended audience and ultimate readership. In the case of yarn, for example, and of sewing thread, records of production and sales are available, but not of consumption: how many skeins of knitting yarn are sold to knit goods manufacturers, and how many to hand knitters? I intend to show that although hedonizing technologies differ markedly from each other and may include multiple technical paths, some generalizations can be articulated regarding patterns of demand for publications about them. I explain in detail my bibliometric methodology in Appendix B and describe its known biases and limitations.

The overall picture constructed from the WorldCat database is that early entrants into the hedonizing marketplace for publications, such as embroidery and hunting, are immediately obvious, as are the late bloomers like sewing and soapmaking. While there are, as we shall see, many differences in the publication trajectories of these subjects, many have elements in common across a broad range of subject headings. There is usually an identifiable point in the development of a hobby or leisure activity at which histories of it, and/or guides for connoisseurs, begin to appear in the literature. The appearance of a serial title on the subject suggests that someone, at least, believes there is a market for topical information on a regular basis; this development may also signal the arrival on the scene of advertisers hoping to makes sales in the leisure market. Some of these serial titles, of course, are not exclusively hobby-oriented; it would in fact be difficult to determine the intended readership of many of them. *Zymurgy*, for example, is written for homebrewers but is sometimes read by professional brewers looking for new ideas. Often, as in the case of the textile crafts, the number of book titles published in a decade levels off as the number of available serial titles increases, suggesting that at some point serials begin to absorb part of the overall market growth in the subject, leaving book publishing with less "market room" for growth.

With plain sewing firmly in the grasp of the apparel industry, and apparel prices at retail falling as a result, home sewing and needlework in Britain and the United States were actively pursued as leisure-time hobbies in the period between the 1870s and the Great Depression. The home was changing rapidly during this period as well, becoming not only cleaner and better lit, but also more congenial in other ways as a recreational workspace. Furniture, for example, had

Central heating and electric lighting have made twentieth-century homes comfortable as leisure workspaces, as this 1920 International Heater advertisement suggests. From International Heater Company, *International Onepipe Heater Makes Your Entire Home Warm, Cozy and Comfortable* (Utica, NY: The Company, 1920).

entered the period of modern comfort at more or less affordable prices, especially when purchased on the installment plan that had been pioneered by the sewing machine.[31] An increasing number of American homes were centrally heated, allowing families to spread their home activities out of the kitchen, which in the early nineteenth century was often the only room heated in winter. Needlework and other publications of this era show families seated in living and dining rooms, with adequate seating and lighting for needlework, reading, card and board games, and other leisure activities.[32]

By the late 1870s, needlework manufacturing and retailing in the United States, Britain, and Europe were well on their way to their modern form. Nearly all yarn and thread types were factory-made and available at affordable prices; the prices of needlework tools and equipment had fallen to levels unimaginable before 1800; and the industries of publishing and distribution permitted national advertising of branded merchandise of reliable quality. Perhaps most importantly, the spread of the sewing machine was everywhere reducing the amount of time required for producing garments and household textiles and the amount of money relative to wages required for purchasing them. The way was finally open for all economic classes to find pleasure and expression in the needle arts. Entrepreneurs at all levels—local, regional, and national—sought to benefit by the opening of these new markets.

New materials for needlework and improvements on familiar threads and yarns were introduced in large numbers between 1870 and 1930 at affordable prices to consumers whose needlework educations included not only traditional instruction by older family members and sometimes formal schooling, but also an increasingly sophisticated needlework press that had begun, by the late nineteenth century, to offer detailed instructions in new techniques as well as patterns for stitches and methods already known to home needleworkers.

Under the combined sponsorship of the household press and the needlework industry, textile media were systematically explored and presented to actual and potential consumers by the most effective means then available: good design, imaginative marketing, and superior materials at popular prices. Magazines, design firms (many of them women-owned), and materials producers offered patterns through the mail and in local retail outlets, which stimulated the purchase of yarn, thread, and fabrics. Designs that were attractive and challenging made excellent advertising for the products required to make up the work, so thread companies published their own patterns, including some submitted by consumers, for work that called for their own line of supplies and notions, much as food manufacturers distributed recipes. They introduced a great variety of new or reintroduced techniques by hiring professional designers, many of whom were also employed by the subscription needlework press. In the United States, ethnic styles and methods were adapted in design publications to American materials. The influence of notions and thread manufacturers was also felt in their significant subsidy of household magazine publishing through advertising. Thus the economic interests of manufacturers and publishers became a much more important factor in needlework design at the end of the nineteenth century than it had been fifty years before, when pattern publishing and yarn and thread production were entirely separate enterprises at the manufacturing level, even though typically merchandised together at retail.

The needlework and household magazines of the first half of the nineteenth century had been priced mainly for the middle class and above. *Godey's*, for example, was $3 a year in 1859, and *Ladies' Home Magazine* $2.25, about what a male teamster or blacksmith earned in a day.[33] Their successor publications of the post-1870s, on the other hand, were able to provide better value at lower prices, mainly because of advertising revenue and because of the falling prices of printing color and photographic illustrations. *Needlecraft*, a popular tabloid-format monthly newspaper published in Augusta, Maine, between 1909 and 1935, was fifty cents a year or five cents a single copy, the same price as a single spool of cotton sewing thread at the same period.[34] The *Star Needlework Journal*, published by the American Thread Company between 1916 and 1925, sold for forty cents a year or ten cents a copy. Both magazines were profusely illustrated and regularly featured the work of popular designers. The *Home Needlework Magazine*, published for Nonotuck Silk by the Florence Publishing Company, contained numerous and attractive color plates every other month for thirty-five cents annually or ten cents a single copy. The *Columbia Book of Yarns*, published by the Philadelphia-based Columbia Yarn Company, was a handsome publication with color covers that sold for fifteen cents a 97–page copy in 1901. By 1918, it was in its nineteenth edition, which ran to 240 pages at the wartime price of twenty-five cents an issue.

Needlework Education and Charity Work

The efforts of schools to enforce arts and crafts as virtuous uses of leisure—or worse, as methods of home production—have often had the opposite effect from what was intended, turning hobby interests into something "a body is obliged to do," a clearly counterproductive result. During the 1960s and 1970s, members of what was then called the National Notions Association (now the National Home Sewing Association) complained that public-school sewing classes created, in far too many cases, lifelong aversions to the art.

The strategy of encouraging sewing, needlework, woodworking, and other crafts by suggesting to students that these arts would, say, cause hair to grow on the palms of their hands did not, apparently, occur to many administrators or curriculum planners. In the late nineteenth and early twentieth centuries, sewing was added to the curriculum in most British schools, and in many American schools as well. Surviving textbooks and sample books testify to the uninspiring content of these courses: hemming, French seams, apron construction, buttonholes, and even "thimble drill."[35] Because home sewing meant less-expensive clothing than ready-to-wear, it did well during the Depression years in the United States. But by 1950 public school students were beginning to show the resistance to plain-work curricula that was the forerunner of the flat home-sewing markets of the 1970s and 1980s (quiltmaking being the exception). Machine knitting has a similar history; the market for home knitting machines has been small and mainly limited to home producers, rather than hobbyists, since the technology was introduced at the turn of the twentieth century.[36]

War and charity needlework, in the early years of the twentieth century, were often "agonized" in a similar manner, by making them a duty and chore. The American traditions of supporting the troops in the field through needlework and home preserving are documented as far back as the Revolutionary War, when Dorothy Dudley wrote in August 1775 that "our hands are soldiers' property now; jellies are to be made, lint to be scraped, bandages to be prepared for waiting wounds. Embroidery is laid aside and spinning takes its place."[37] We do, however, have some accounts of British and American women enjoying not only the work itself in these contexts, but also of the knitting of socks and putting up of preserves for soldiers being enjoyed as group social activities. Some also mention the satisfaction of being able to do something to help; women usually could not fight, and few could leave their families to serve as nurses and relief workers, but many could and did make socks, caps, sewing kits, and other gear for Civil War and World War I soldiers, and garments of all kinds for European refugees in both world wars.[38]

Nevertheless, charity needlework already had, by the twentieth century, long traditions of tedium, obligation, and agonization by religious groups and eleemosynary organizations. As we have seen, plain sewing and knitting for the poor and for hospitals were imposed on nuns and other women as virtuous chores at least as early as the Middle Ages, and the Biblical reference to Dorcas suggests that the tradition is even older. Nineteenth-century sources, in fact, sometimes refer to charity needlework as "Dorcas work."[39] Somewhat less tedious, but subject to similar social pressures from clergy and altar guild members, was the production of ornaments for religious buildings, including altar cloths, kneelers, and banners for churches and Torah covers for synagogues, as well as production for charity fairs and bazaars. This tradition remains in a grey area between obligation and pleasure.[40]

The Fashion for Ethnicity

In the United States, the fashion for ethnic designs, led by Thérèse de Dillmont and her contemporaries in Europe in the late nineteenth century, became a veritable craze in the first three decades of the twentieth. Needleworkers borrowed ideas from one technique and applied it to another, as in the cases of mosaic embroidery, an innovative use of surface embroidery techniques applied to counted fabrics, and of Mrs. F. W. Kettelle's designs for embroidery on filet crochet, many of which had the appearance of the popular ethnic Norwegian embroidery technique known as hardanger, a technique of embroidery over drawn threads.[41]

This Norwegian technique was only one of many European ethnic styles and techniques that became part of American mainstream needlework at the turn of the twentieth century, appearing in respectable WASP parlors and dining rooms, in most cases long before the originators of these crafts were invited there.[42] Danish hedebo and the counted-thread technique on huckaback fabric called "Swedish weaving" were both well represented in needlework magazines.[43] British and German styles, of course, had been accepted in the United States since patterns from Europe began to be imported in the nineteenth century; and the nearly ubiquitous presence of Americans from these ethnic backgrounds, along with their growing respectability and buying power, assisted in the assimilation of their needlework styles to the American mainstream after the 1870s. Flora Klickmann's 1915 *Cult of the Needle,* for example, an anthology of patterns originally published in *Girls' Own Paper* and *Woman's Magazine,* included needlework designs from Hungary, the Greek island of Rhodes, the Catalan province of Spain, Russia, Ireland, Italy, Denmark, Norway, Belgium, Bulgaria, Bohemia, and Sweden.[44]

Irish, Italian, Armenian, and Hispanic ethnic needlework styles and techniques were examples of widespread acceptance of imported textile handcraft methods by Americans who were perhaps not yet entirely ready to embrace diversity in their personal and professional lives.[45] The first two cases, Ireland and Italy, both had long traditions of artistic achievement that Western Europeans and Anglo-Americans had become accustomed to regarding, somewhat paradoxically, as aesthetic phenomena distinct from the peoples and cultures that made them.[46] Italian patterns were appearing in the American and British needlework press by 1889 and were regularly featured through the 1930s.[47]

The sympathy and aid efforts of the international community for the victims of the Armenian genocide of 1915–23 helped bring ethnic Armenian needlework into fashion in the United States and Britain. Hispanic needlework seems to have been regarded simply as exotic and interesting, much as Filipino and Chinese styles were. Novelty was nearly always a selling point with hobby needleworkers, except for the members of an affluent elite who cherished fantasies of a lost paradise of fine stitchery. Of these well-heeled traditional revivalists, I shall have more to say in the following section, Nativist Backlash.

Most American hobby needleworkers of the early twentieth century, however, wanted new and challenging needlework ideas for traditional uses—clothing, gifts, and household accessories such as doilies and antimacassars—as well as for a few new ones, such as potholders and covers for hot-water and condiment bottles. The fancy work in these items was entirely superogatory, as it is possible to make a perfectly serviceable child's cap, table cover, pillowcase, or kitchen apron entirely of plain work done on a sewing machine. Indeed, by this period it was not unusual for artisans to buy cut-out or even completely finished articles in usable form, but with embroidery patterns stamped on them, so that the embroiderer need not bother with the plain work of assembly, proceeding directly to the pleasure of embellishment.[48] The lace and/or embroidered edgings for towels, pillowcases, sheets, handkerchiefs, and underwear that were nearly ubiquitous in needlework and household magazines between 1880 and 1930 served no useful purpose; they were made to delight, both in the doing and in the using. For this purpose, it apparently did not matter what the nominal ethnicity of the pattern was: American hobby needleworkers were delighted to try their hands at techniques from Ireland, Japan, Romania, the Canary Islands, Greece, Yugoslavia, or anywhere else; and the market was sufficiently large to send designers far afield in search of them.[49]

Few of the interpreters of these imported techniques were actually members of the ethnic groups whose work they popularized, but were instead collectors and interpreters of global needlework, much as their Austrian-Alsatian predecessor Thérèse de Dillmont had been. There were exceptions, of course—for instance, Carmela Testa's Italian filet, cutwork, and Assisi and Florentine embroidery patterns in the pages of *Variety* in the 1920s; and Rita Garcia's patterns for colado Philippine punchwork embroidery in *Needlecraft* in the teens.[50] U.S.- and British-born designers, however, had the lion's share of space in the needlework pattern press.[51] Anne Champe Orr, for example, one of the many non-Irish designers who made Irish crochet a household word in the early twentieth century, was a Nashville, Tennessee, entrepreneur with a traditional art education. In her lifetime, she published nearly a hundred volumes of needlework patterns.[52] As Anne Champe, she published pattern books herself, designed for thread companies. After marrying wholesale merchant J. Hunter Orr in 1919, she became the needlework editor of *Good Housekeeping*, which was still receiving fan letters addressed to Orr as late as 1977, twenty years after her death.[53]

Irish crochet, with its elegant three-dimensional floral and shamrock "flower" and picoted brides (connecting net), was so popular after the turn of the century that it was imitated in other techniques, such as tatting—an irony, of course, since Irish crochet is thought by some to have been itself an adaptation to Hibernian conditions of the bobbin laces made in Belgium, much as Tilsit cheese was originally intended as a Danish knockoff of Swiss Emmenthaler.[54] Irish crocheted lace, mainly in white, cream, and ecru, was used for neckwear, especially collars and jabots, cuffs, doilies, jewelry, blouses, caps, and, improbably, even shoes. Thread companies loved the technique, not only because it was popular but because it was more thread-intensive than, for example, filet crochet, which was a flatter, more two-dimensional medium.[55] Designers loved its versatility and glamorously "dainty" appearance, and many well-known needlework designers, including Anne Champe Orr, published patterns in the Irish ethnic style.[56] Mountmellick embroidery, a sophisticated technique of raised whitework thought to have originated in Ireland, was also popular in the United States at the turn of the century.[57]

The Armenian catastrophe of 1915–23, in which close to a million Armenians died in or on their way out of their homeland, brought many Armenians to the United States, including a number of accomplished needlewomen, who set out not only to earn their own livings but also to aid their starving and displaced compatriots with the proceeds of their handcraft.[58] Armenian lace became a fad both in its original form as a needle-made mesh and reinterpreted in other forms such as crochet. The Armenian designer Nouvart Tashjian's patterns were published in *Plain and Fancy Needlework* in 1917 and in *Modern Priscilla* in 1922. *Plain and Fancy* placed an advertisement for Tashjian's lace patterns on the back cover of the April 1917 issue, telling readers that "Miss Tashjian will teach you how to make this and at the same time you will help the Armenian Relief Fund."[59]

Nativist Backlash

Not everyone shared this enthusiasm for ethnicity. The embroiderer Martha Genung Stearns (1866–1972), quoted earlier in the opinion that the twentieth century had contributed "almost nothing to American needlework," considered that "the end of needlework" had come with the mechanization of textile production and identified this melancholy demise as dating from the invention of the cotton gin. The "triple fellowship" of spinning, weaving, and embroidery was "broken" by the introduction of mass-produced fabrics and yarns, she sententiously opined. "And with the breaking off of the thread, we surrendered that fellowship, and the love of wrought decoration which was bound up with the fabric loosened and fell away." For the twentieth-century flowering of crochet, tatting, knitting, and ethnic needlework of all types, Stearns had nothing but contempt: "We have passed successively through the Berlin woolwork period with its slippers decorated with red and purple flowers, and cross-stitched hairy dogs carrying their master's cane in their jaws worked upon 'catch-alls,' the Biedermayer and Battenberg period, Mexican drawn-work, Swedish hardanger and hedebo, Italian cutwork, French eyelets and satin-stitch; we have decorated our intimate garments with Hamburg [lace] and tatting, crochet lace in wool for winter flannel petticoats and in cotton for summer nainsook ones, and the final enormity of 'lazy-daisy' [an embroidery stitch], all of which left us poorer in design and more careless in execution than before."[60] One is tempted to speculate that New England Republican bluebloods like Stearns may have resented the democratization of arts that had been the exclusive preserve of an affluent and leisured minority before the nineteenth century.[61] The snobbery, arrogance, and scorn for popular crafts

Stearns expresses had roots in elite aesthetic movements that had their heydays long before Stearns wrote in 1963.

Not all of the proponents of these movements, however, were as overtly xenophobic as Stearns; some were mainly interested in preserving crafts they feared were threatened by industrialization. Needle and bobbin-made laces, handweaving, hooked rugs, embroidery, traditional costume, and eventually quiltmaking—but not, of course, the "déclassé" arts of crochet, tatting and knitting—were the topics of a number of books and museum exhibitions between 1880 and World War II. Needlework was exhibited at the Victoria and Albert Museum in London, the Metropolitan Museum of Art, the Newark Museum, and the Brooklyn Museum in the United States, the Bartholdi Fair in New York City, the Centennial Fair in Philadelphia, the World's Columbian Exposition in Chicago, and many other prestigious venues.[62] Much of this documentary activity was directed at collectors and prospective collectors of these supposedly lost or endangered textile arts; the rest consisted of encouragement to revive these arts by the publication of patterns and instructions. Both of these types of literature—connoisseurship and revival artisanship—although biased in ways we have already discussed, were important contributions to the historical documentation of artisan textiles.[63]

Part of the background of this relatively new interest in textiles as art was the general need, in Europe and the United States, for a convincing national image with a historical narrative that could be used as a basis for what has come to be known as "civic religion." In the United States, we are familiar with the rage for Colonial Revival of the early years of the twentieth century, but other nations were developing counterparts at about the same time. Britain was, by 1900, already engaged in the preservation and mythologizing of its past that was, a hundred years later, to make it such a successful tourist destination that its Lord Chancellor complained in the 1980s that the country had begun to resemble "a vast historical theme park rather than a modern industrial nation."[64] Other European nations, both before and after World War II, launched programs of collecting and preserving their national artifactual traditions, including needlework and other traditional objects and patterns.[65]

One of the earliest of these "back to the future" trends to influence needlework was the British movement to bring higher artistic standards to industrial design. Embroidery, lace, and fine weaving were exhibited in London in 1851 and were shown in *The Art Journal Illustrated Catalogue: The Industry of All Nations.*[66] The Arts and Crafts movement in Britain and the United States was explicitly aimed at reviving and perpetuating traditional crafts in an industrial era, although this movement retained the element of aristocratic snobbery in both countries. In the closing decades of the nineteenth century, figures like the Englishmen Edward Burne-Jones (1833–98), John Ruskin (1819–1900), and William Morris (1834–96) and the American Louis Tiffany (1848–1933) saw themselves as reestablishing the value and dignity of heritage artisanship, champions of skilled manual labor in the Age of the Machine.[67]

In Britain, the South Kensington School worked with artists Burne-Jones, Morris, and Walter Crane (1845–1915) in the restoration of traditions of artisan lace and embroidery to what were perceived as their former professional glories, a movement that eventually brought the UK Embroiderers' Guild into existence in 1906. Another aesthetic enterprise, the *Werkbund*, tried, with very limited success, to bring labor and art together in Germany.[68] In the United States, the Society of Decorative Art, with the encouragement of Louis Comfort Tiffany and embroiderer Candace Wheeler, taught advanced silk and wool embroidery to affluent, white, well-educated modern women; and the Boston Society of Arts and Crafts, founded in 1897, labored to revive crewelwork.[69] American needleworkers who wanted to identify with the British leisured elite could buy pattern books from a New York and London organization styling itself "The Royal Society." Despite the aristocratic name, most of the patterns were for conspicuously ordinary crochet and knitting projects. The connection with royalty was obscure.[70]

Meanwhile, two New York–trained artists, Margaret C. Whiting and Ellen Miller, described by Martha Stearns as "descendants of Colonial families," were apparently so deeply impressed both by the Arts and Crafts movement values and by the Colonial Revival that they established an enterprise employing local Massachusetts village women to reproduce blue and white crewel embroidery in what they took to be the colonial American style and selling the product in the "carriage trade" market for colonial-style home furnishings.[71] The Deerfield Society of Blue and White Needlework, founded in 1896, did not survive past 1925, possibly because it rested too precariously on the economic boundary between hobby and labor, but it was characteristic of the early-twentieth-century Colonial Revival movement.[72]

All things colonial were thought, in that period, to be good and beautiful, and when they were not, were reinterpreted by historical spin doctors like Josephine Wheelright Rust of the Wakefield National Memorial Association, who, upon discovering in the 1920s that George Washington was born in a small, quite unassuming two-story house on Virginia's Pope's Creek, proceeded to build him a "birthplace" in 1931–32 better suited to his role as the Father of His Country.[73] The *Index of American Design*, a Works Progress Administration project of the

1930s, contributed to and documented the fascination with colonial decorative arts.[74] A spate of handcraft and home decorating books and articles celebrated and encouraged revival of colonial textile and other crafts.[75] The Whitman Candy Company of Philadelphia, faced in the 1930s with little demand for its product but with the certainty of future ruin if it laid off its highly trained and experienced staff of chocolate-dippers, sent them off into the countryside to collect samplers in honor of the company's "sampler" box. This valuable legacy of the Colonial Revival tradition of connoisseurship remains available to researchers as part of the vast textile collection of the Philadelphia Museum of Art.[76] Quilts, formerly scorned as plain work, were rehabilitated by collector-quiltmakers Ruth Finley, Rose Kretsinger, Carrie Hall, and others as early American art objects in the 1930s, setting in motion the forces that were to result in the late-twentieth-century craze for hobby quilting and patchwork, to which we will return in the next chapter.[77]

The movers and shakers of Colonial Revival, as Laurel Thacher Ulrich has pointed out, focused much of their attention on archaizing American kitchens, an environment which had been undergoing significant transformations of its own.[78] It is to the changing role of the kitchen as a leisure workspace that we turn next.

Home, Garden, and Garage as Leisure Workspaces
Kitchen Hobbies

Like needlework, cooking was an early beneficiary of printing technology, with more than seventeen hundred cookbook titles published before 1800, including one serial (see Appendix B). By the end of the first decade of the nineteenth century, cooking had several serials, which were mainly aimed at professionals such as confectioners and chefs. Several hundred titles were published every decade through the nineteenth century until the 1880s, when cookbook publishing expanded abruptly to more than fifteen hundred titles in that decade alone. In the following decade, the total number of titles doubled to nearly three thousand, even as the number of prepackaged foods available to consumers increased steadily. Urbanization may have played a role; young people moving to cities for employment may not have had the access to the narrated demonstrations that had formed the basis of culinary education for their grandparents.

The cookbook trend continued right through the twentieth century; the less Americans actually needed to cook—and by 2000, very few of us really needed to do much more than push buttons on a microwave—the more cookbooks we bought, as the technology of cooking from scratch was hedonized into a hobby, just as needlework had been before it. *Gourmet* magazine began publishing in 1941, and *Bon Appétit* in 1956. Between 1991 and 2000, a record 48,000 cookbooks were published, and 272 new cooking serials began publication. During the same period, the number of meals prepared in the average American home declined steadily, from about 20 per week in 1950 to 14.4 in 2000.[79] The "preparation" of several of those 14.4 meals typically consists of making a sandwich, pulling a container of yogurt out of the refrigerator, or pouring cereal and milk into a bowl. In 2002, cookbook sales in the United States amounted to something on the order of $433 million, and $6.6 billion went into kitchen makeovers at an average of $43,800 each.[80] Clearly, those of us who are still cooking are doing so because we like it; the author's parents, who have always hated working in the kitchen, now buy nearly all their comestibles fully prepared at Whole Foods. We hedonized cooks, on the other hand, order professional-quality All-Clad cookware from Chef's Catalog, glory in our alphabetized whole-wall spice racks, and steam up the windows while we prepare Peruvian-style braised pork shoulders simmered with tiny blue potatoes.

As we have seen, the making of preserves, jams, candies, and beverages at home began to hedonize from their roots in still-room tasks in the early modern period, encompassing home brewing and the making of cider and wine as hobbies for the affluent by the seventeenth century.[81] As in the case of cooking, the expansion in print resources for brewing and canning between the sixteenth and the nineteenth centuries reflected the transition from information transfer via in-person demonstration, walking readers through preparation tasks, often with illustrations and step-by-step instructions.[82] In the nineteenth century, most kitchen, still-room, and brewhouse arts developed separate professional literatures for production processes, as in the cases of beer, vinegar, and canned food production.[83] But amateur manuals continued to appear, at different rates in different crafts; as in the case of needlework, the process of hedonization was deeply nuanced within general categories.

Brewing has, of course, a much smaller literature than cooking, but the growth of printed works and of amateurs' manuals on the subject is almost as impressive on its own scale. Only 139 works on brewing are documented in World-Cat before 1800, but publications on this subject grew in the nineteenth century at the rate of 35 to 70 per decade until—like cooking—by the 1880s, more than a hundred brewing titles were published, including the establishment of fifteen serial titles and several histories and connoisseur's guides, the second and third derivatives of hedonizing technologies. Breweriana, the lore and gear of beer, also

Gender matters: traditionally, home brewing was a chore for women, but some "gentlemen" chose it as a hobby. This grim-faced brewer-by-necessity, a "Mrs. Vernon" who owned a country inn, was photographed in her brewhouse at Cheshire, England, on 7 March 1931. Modern brewing technology would soon make this type of home beermaking obsolete for production, but increasingly popular as a hobby. © Hulton-Deutsch Collection / Corbis, reprinted by permission.

made its first appearance at this time as a collecting specialty. Again like cooking, the number of titles doubled in the following decade, but unlike books on kitchen arts, those of the brewhouse fluctuated in popularity in the following years, in part because of the Volstead Act. By the 1930s, however, brewing titles were being published at the rate of more than two hundred per decade, and forty serial titles began publication. The real explosion, however, consistent with the hypothesis that much of this growth was driven by the hobby of homebrewing, was between the 1960s and the 1970s, when the number of titles rose from 278 to 501. Between 1990 and 2000, nearly a thousand brewing monograph titles were published, and almost a hundred brewing serials are listed in WorldCat as having begun in the 1990s.

Two skills associated with confectioners and bakers—candymaking and cake decorating—have quite different historical patterns of hedonization. One would not guess from visiting A. C. Moore or Michael's that these crafts, the tools and materials for which are similar and typically displayed near each other in the stores, have been hedonized at such dramatically different rates over time. As we saw in Chapter 2, candymaking originated in the still-room tasks of preserving flowers, fruits, and herbs or their essences in sugar or honey, for use in the winter months as medications. As sugar became more widely available and less expensive with the spread of colonialism and sugar slavery, the confectionary arts expanded, and became affordable as leisure hobbies for "ladies" in Britain and Europe and, somewhat later, in the American colonies.[84] In Britain and the United States, candymaking became a mechanized industry in the mid-nineteenth century, acquiring in the process the kind of technical literature we have already seen in the cases of brewing and canning.[85] The availability of relatively inexpensive testing apparatus like candy and brewing thermometers, which began to be mass-produced in the nineteenth century, may have contributed to the hobby trend in candymaking and brewing, and possibly to that of home preserving as well.[86]

The hobby and juvenile-recreation literature continued to burgeon, however, in parallel with professional publications, fed in part by recipes from the manufacturers' test kitchens of products like sweeteners, milk products, flavorings and other foods, including Karo Syrup, Hood's Sarsparilla, Knox Gelatin, Union Starch, Price Flavoring Extract, Walter Baker Chocolate, and Pet Milk.[87] Between 1888 and 1990 more than a hundred titles were published in English on home candy-making. Like needlework patterns, candy recipes were used to sell other products, such as Lydia Pinkham's medicine and cooking gas, the latter probably based on the greater ease of controlling temperature-critical processes, such as caramelization, on gas or electric appliances than on stoves that used coal or wood as fuel. As in the case of needlework, candymaking had a small literature of home production for bazaars and sometimes for sale.[88] A few candy cookbooks, most of them by state agricultural extension services, were published with the aim of providing recipes for less expensive confections made at home, but by the 1950s nearly all amateur handbooks for confectionery were marketed in the rhetoric of pleasure.[89] In the decade following 1990, another thirty titles appeared, with every indication that the rate of publication was continuing to increase. Candy-making artifacts had become "collectible" by the 1970s.[90]

Cake decorating, on the other hand, despite its reliance on much the same types of ingredients and skills, had only a small, almost entirely professional literature from its origin, in an 1815 French work, until after World War II.[91] Between the 1890s and 1950, a modest number of monographs and only a handful of serials were published about cake decorating.[92] In the 1920s, for example, twenty-two books on cake-decorating were published, and no serials are known to have begun publication during this period. Between 1941 and 1950 (inclusive) only eleven titles, all monographs, are known to have reached print. But in the 1950s, the craft surged into the hobby market, with sixty-one titles, most of them expressed in the rhetoric of pleasure and creativity.[93] From 1971 to 1980, cake decorating titles overall numbered 127 and a serial began publication; by 1990 another 302 titles had appeared, among them three more serials, more than doubling over the previous decade. At this writing, hobby publications on cake decorating per year outnumber those on candymaking by a full order of magnitude.

Why did cake decorating get a slower start, then surge past candymaking in the second half of the twentieth century? I cannot even begin to explain all of the complexities in these evolving markets, but one can speculate that the role of the oven in baking cakes, with its much larger space and fuel requirements than the stove burners used for most candymaking, may have held back the growth of hobby cake decorating. Baking in the summer, of course, has historically been a sweaty and arduous chore, and this is unlikely to have changed before air conditioning become affordable, although the electric fan, introduced early in the twentieth century, may have been a help. The steam-bath aspect of most summer culinary tasks may have been a factor in the early development of home preserving as a hobby because it was possible to preserve and can in summer kitchens that did not have ovens, and summer kitchens, where present in houses, were usually located at some distance from living and sleeping areas.[94]

Two other late-bloomers among kitchen hobbies are the making of soap and candles. The documented literature of candlemaking is tiny compared with most of the literatures we have already examined. Eight titles, all of them by or about professional chandlers, were published before 1800. Between 1800 and 1970, only 28 more titles worldwide are documented in WorldCat, including one serial, *American Candlemaker*, which began publication at the turn of the century. One history of medieval candlemaking appeared in the 1920s, in German, but no clearly identified amateurs' manuals are known to have been published before 1971, although the topic was sometimes included in housekeeping manuals as early as the eighteenth century. In the 1970s, however, the baby boomer generation took home chandlery to its capacious collective bosom, along with two other crafts involving wax, tie-dyeing and batik. Fifty-eight titles on making candles at home were published between 1971 and 1980, only 13 in the decade following, but 97 in the 1990s, contributing, no doubt, to that decade's notable increase in the number of candle fires in American dwellings.[95]

The career of home soapmaking, counterintuitively, has a somewhat more glamorous history. A WorldCat search for the timeframe "early works to 1800" yields a respectable thirty titles, though more of these are about duties, taxes, and production standards for commercially produced soap than about making it at home. Considered by some to be the world's first manufactured chemical product, soap in the marketplace was an early target for regulation of quality. All soap-making serial titles before 1990 documented in WorldCat are unambiguously published for commercial producers.

The first freestanding amateurs' manuals, as distinct from the subject's routine inclusion in larger works on housekeeping, appeared in the first two decades of the nineteenth century and were addressed to quality-conscious "gentlemen and ladies."[96] Two more manuals appeared in the 1880s, published by manufacturers of lye, but the titles of both were expressed in the rhetoric of thrift, not pleasure, a theme that was to recur in the history of this craft (much as it had in candymaking) with the publication of soapmaking instruction booklets by agricultural extension services between 1920 and 1960 in the United States.[97] Lye manufacturers, as we have seen, were the other main source of amateurs' manuals on soapmaking in America before 1981, publishing instructions in magazines, in brochures, and on the lye cans themselves.[98] In the 1970s, the boomers who were turning their hands to traditional arts and publishing guides to them, such as the Foxfire books, adopted soapmaking as part of the "back to nature" component of leisure artisanry.

A commercial history of soapmaking had appeared in the 1890s, and three more appeared in the 1930s, but the collecting of soap advertising, gear, dishes, wrappers, and the like had to wait for the art to gain popularity as a leisure hobby activity in the 1980s, when 67 titles were published. In the decade following, 111 titles were published, nearly all of them for amateurs, and 51 serials began publication.

It is amusing to contemplate how a harried American homemaker of, say, 1830, would have regarded our interest in the making of soap and candles as leisure activities. Like most other hedonizing technologies, these tasks were regarded as work until the manufactured product became so inexpensive that the time to make it cost more than it was worth, as in the case of soap, or until the product itself was not merely technologically obsolete but antiquated and slightly dangerous to use, as in the case of candles. Although home cheesemaking has not caught on the same way that, for example, home preserving has, it is nonetheless accumulating a hobby literature that closely resembles the other examples we have examined here.[99] One can only wonder whether the future will see the hedonization of such tasks as washing dishes by hand, or manual cleaning of ovens.

As kitchen activities, either as work or as leisure, moved away from the open hearth into the greater convenience, cleanliness, and efficiency of stoves between the late eighteenth and early twentieth centuries, the fireplace became by the mid-twentieth century a hedonizing appliance, usually located in the living room, where it became the venue for such cooking-for-pleasure activities as roasting chestnuts, popping corn, and toasting marshmallows, when it was used for any purpose other than enjoyment of the sight, smell, and sound of wood oxidizing to no practical purpose.[100]

Outdoor Cooking

Cooking real food over wood and charcoal was on its way out of the home kitchen by the 1920s, but after World War II, it reappeared in the back yard.[101] Outdoor cooking was, of course, for millennia associated with hunting and fishing expeditions, and with military travel, and was by the nineteenth century, at the very latest, recognized as a pleasure in itself, as well as a component of the hedonization of the outdoors, a topic we shall explore later in this chapter. While aristocratic campers and hunters brought cooks to do the work, less affluent outdoorsmen had wives to do the cooking and cleaning chores on camping trips as well as at home.[102] That some women did not relish this role is evident from Elizabeth Von Arnim's portrayal of a baron and his family and friends on a camping expedition in her 1930 novel *The Caravaners*. Von Arnim's narrator, Baron Otto, is shocked and horrified by the amount of labor required by what is intended as a form of leisure and cannot understand why he should be required to do any of it: "At every camp there is nothing but work, — and oh my friends, such work! Work undreamed of in your ordered lives, and nothing, nothing but it, for must you not eat? And when you have eaten, without the least pause, the least interval for the meditations so good after meals, there begins that frightful and accursed form of activity, most frightful and accursed of all known forms, the washing up. How it came about that it was not from the first left to the women I cannot understand; they are fitted by nature for such labour, and do not feel it; but I, being in a minority, was powerless to interfere. Nor did I always succeed in evading it."[103]

When women were not included in the party, and sometimes even if they were, some men and boys learned to enjoy doing the cooking themselves; the first edition of the *Boy Scouts Handbook* (1911) included some recipes and cooking instructions.[104] Girl Scouts, however, received much more elaborate instructions in outdoor cooking, including, by 1946, several pages of drawings and diagrams for the construction of a "primitive kitchen," including utensils made by

The labor-intensivity of camping includes cooking, cleaning, washing dishes, child care, airing of bedding, laundry, and many other tasks made exceptionally inconvenient by the outdoor venue. This family, roughing it in the White Mountains National Forest, New Hampshire, in April 1962, even had to change and wash diapers. Photo by Dick Smith © Corbis, reprinted by permission.

hand in "troop meetings with the aid of a good pair of tin shears, a Scout knife, and a bit of skill, imagination and ingenuity."[105]

The earliest references I have found to the modern Western hedonized practice of barbecuing, with its tenaciously male associations, are to outdoor political, religious, and social occasions, primarily, but not exclusively, in the American South and West, in the eighteenth century. Washington Irving (1783–1859) reported on an event he attended about 1809, "a great 'barbecue,' a kind of festivity or carouse much practised in Merryland [*sic*]."[106] An anonymous Virginian wrote in 1860 that barbecues were being held in his state at the time of the Revolutionary War, and there is evidence that the same was true of Kentucky.[107] In 1851, Senator Solomon Downs gave a speech on the Compromise of 1850 at a barbecue in Louisiana; we have a document from three years later announcing an Independence Day barbecue in honor of the arrival of the steam railroad in Montgomery County, Virginia, and another from 1855, with the text of a Louisiana politician's speech presented at the Houma barbecue in Terrebone Parish.[108] Veterans of both the Union and the Confederate armies held reunion barbecues after the war, a practice that must have recalled and reinforced the masculine traditions of military camp life.[109]

In the late nineteenth century, barbecues were the subject, as well as the venue in many cases, of songs and minstrelsy, including both nostalgic tributes to Southern cooking and hospitality, and what to the modern reader are truly cringe-worthy "comic" sketches of African-American outdoor social events.[110] Barbecuing is documented in New England by 1815, and in California by 1881; clearly, by the end of the nineteenth century, the outdoor grill was on its way into the mainstream of American life, although not yet into individual backyards.[111] As every firefighter knows, barbecues are a traditional fundraising method for volunteer fire departments, which, while now partly gender-integrated, remain largely male institutions; churches, too, often raise money by selling tickets to "picnic socials" and cookouts staffed by usually male chefs wielding impressive meat forks and barbecue tongs.[112]

Sunset Magazine published directions for building a backyard barbecue in 1931, and an elaborate do-it-yourself manual and cookbook in 1945, with direc-tions even on how to do the metalwork and masonry required to build both per-manently installed and portable indoor and outdoor barbecue grills.[113] The de-tailed explanations of how to build and use these appliances suggests that they were not reliably available at retail in this period. In the 1945 edition, sixty-six pages were devoted to the design, building, and accessorizing of barbecue grills and ovens, and only thirty to recipes. The formula must have been a success be-cause the book was in its twelfth printing by April 1948.

Ford Motors contributed to the barbecue trend by promoting its grills and charcoal briquets in the 1930s, and producers of grill-worthy meats, sauces, and other products were not slow to publish recipes and instructions, wooing the leisure market just as their counterparts in needlework and candymaking had done.[114] In the postwar period, barbecue cookbooks came thick and fast after *Sunset*'s entry into the field, with at least sixty-two titles between 1945 and 1960. Barbecue in the second half of the twentieth century entered the social dining canon in Australia, South Africa, and the United Kingdom as well as Canada.[115]

Although some titles were written by women for the usual female readership of other types of cookbooks, the tradition of barbecue books written by and for men was already well established by the 1960s and remains alive and well in hairy-chested titles like *Boy Gets Grill; A Man, a Can, a Grill;* and *Secrets of Cave-man Cooking for the Modern Caveman* and in more sophisticated forms of culinary

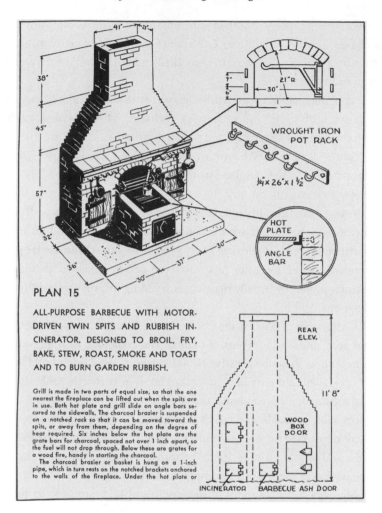

PLAN 15

ALL-PURPOSE BARBECUE WITH MOTOR-
DRIVEN TWIN SPITS AND RUBBISH IN-
CINERATOR. DESIGNED TO BROIL, FRY,
BAKE, STEW, ROAST, SMOKE AND TOAST
AND TO BURN GARDEN RUBBISH.

Grill is made in two parts of equal size, so that the one
nearest the fireplace can be lifted out when the spits are
in use. Both hot plate and grill slide on angle bars secured to the sidewalls. The charcoal brazier is suspended
on a notched rack so that it can be moved toward the
spits, or away from them, depending on the degree of
heat required. Six inches below the hot plate are the
grate bars for charcoal, spaced not over 1 inch apart, so
the fuel will not drop through. Below these are grates for
a wood fire, handy in starting the charcoal.
   The charcoal brazier or basket is hung on a 1-inch
pipe, which in turn rests on the notched brackets anchored
to the walls of the fireplace. Under the hot plate or

Barbecue construction for the serious multitasking do-it-yourselfer of 1945, from *Sunset Magazine*. Building your own barbecue was just another exciting option for postwar hedonizers of outdoor cooking. The construction of yard features like this one was typically a "guy" hobby; the distaff side could contribute hand-crocheted potholders. © Sunset Books, reprinted by permission.

machismo like *Patio Daddy-O* and *Big Daddy's Zubba Bubba BBQ.*[116] Women have traditionally been disposed to be supportive of male outdoor cookery in this context, even making handmade barbecue mitts and aprons for it, but also to be either openly or secretly amused by the ritual aspect of what for women who are cooking indoors is considerably more matter-of-fact. Cecelia Ahern's female protagonist Holly, a widow, observes a man barbecuing in the 2004 novel *PS, I Love*

You: "Holly . . . giggled at the thought. Her dad got so excited when they had barbecues; he took the whole thing so seriously and stood by the barbecue constantly while watching over his wonderful creations. Gerry had been like that too. What was it with men and barbecues?"[117]

One of the male characters in Jonathan Franzen's 2001 novel *The Corrections* is gradually losing interest in nearly every aspect of his life until he encounters the concept of mixed grill, which, for awhile, becomes a kind of emotional life-preserver of pleasure: "It was such a treat that he began to do his own mixed grills at home . . . braving all but the foulest weather on the deck, and loving it. He did partridge breasts, chicken livers, filets mignons, and Mexican-flavored turkey sausage. He did zucchini and red peppers. He did eggplant, yellow peppers, baby lamb chops, Italian sausage. He came up with a wonderful bratwurst–rib eye—bok choy combo. He loved it and loved it and loved it and then all at once he didn't."[118] As Franzen's account suggests, modern barbecue has moved well beyond the roasted pigs and bean pots of the nineteenth century, reaching a level of maturity in the twenty-first century that encompasses vegetarian and even vegan diets. It has also, predictably, acquired a connoisseurship of grills, both factory-made and artisan produced.[119]

Garage and Workshop Hobbies

The hobby craft of modelmaking, a primarily male leisure activity in Britain and America since the late nineteenth century, had its origins, like many other artisan hobbies, in a tradition of miniature construction for practical purposes, mainly education, experimentation, patent demonstration, and administration of complex systems over larger geographical areas than could be overseen by eye. Architectural models could be of any of these types. In the first category were orreries, models of the solar system, used to explain the motions of planets around the sun, and machine models exhibited to inform exhibition visitors, a practice documented as early as 1683 in France and probably much older and more widespread.[120] The "Cosmorama" of a Spanish slave ship exhibited in London in 1840 for the purpose of exposing the horrors of the Middle Passage was also of this educational type, as are anatomical models of all periods.[121] Experimental models, including those of ships, vehicles, and mechanisms of every sort, are well documented in the publications of the Royal Society, the Franklin Institute, and elsewhere from the late seventeenth century forward, as are patent models, a category with which experimental models overlap.[122] Of the last type, administrative models such as the relief model of Paris built in the mid-eighteenth century by one de Quoy, an engineer in the service of Louis XV, "after twenty-two years of a most and laborious and painful Application," is an example, as is Eduardo Jose de Moraes' *Plano Geral da Viação Ferrea da Provincia do Rio Grande do Sul*, presented to the Brazilian imperial governor in 1882.[123] However likely it may be that the governor yielded to the temptation to play with his lavish and expensive model train layout, this and the other types of models described here were devised with serious rather than playful intent; they were tools, not toys.

The fun and frivolous elements of modelmaking, however, were perfectly evident by the 1860s, when an already-gendered juvenile literature began to appear in Britain and the United States, with titles like *Boys' Pump Book: Showing How to Make Several Kinds of Miniature Pumps and a Fire Engine* and *The Boy's Own Book of Boats*, on making sailboat models.[124] Adult males were evidently already well on their way to model mania as well: the London exhibition of 1862 featured a model of Lincoln Cathedral built by British agricultural laborer James Anderton of "one million eight hundred old bottle corks," and by the 1880s, amateurs' manuals like *Model Yachts and Boats* were being published for an adult readership.[125]

By 1890, *Scientific American* could confidently assert that "many young men ride a mechanical hobby, and are often building experimental machines, and making 'young' steam engines." Electrical models were already popular among these hobbyists as well.[126] Model boat–making was so popular in Britain by 1901 that its enthusiasts reportedly supported a weekly journal, *Work*, devoted exclusively to "the subject of building, rigging and sailing model boats."[127] Much time and engineering skill were devoted to methods of making machine models—including steam locomotives, automobiles and ships—work as their full-size counterparts did.[128] By the early years of the twentieth century, leisure modelmaking not only had its own literature and its own marketplace for tools and materials but had picked up its third derivative as well: collecting, connoisseurship, and museum exhibitions.[129]

That amateur modelmakers did not regard their work as the "painful Application" attributed to de Quoy in the eighteenth century is evident from a 1903 account in *Scientific American* of the "private workshop" of one Dr. Frank H. Brandow, "president of the Berkshire Automobile Club, of Pittsfield, Mass.," who produced "during leisure hours" a 65–pound model of a locomotive which

> differs from the full-sized engine only in being fitted with a brake that works by steam instead of air pressure. Steam for the brake is supplied from a tank just above the forward truck, the tank generally used for air pressure in the Westinghouse brake system. The engine is built of brass, bronze, silver, copper, nickel plate, cast

iron, aluminium, and gold plate. There is no wood-work in its construction. The tank is built of burnished copper riveted in the usual way . . . The holes drilled in the boiler at the side and the bottom are used for draft for the alcohol burners used to make steam, it being impossible to generate steam in so small a boiler with flues on account of lack of draft.[130]

Dr. Brandon's creation shows "skill and ingenuity" and a passion for his hobby sufficient to spend a year and a half crafting what is essentially a toy for grown-ups. It also shows clear evidence of a home workshop supplied with tools and materials that would have done credit to a university engineering laboratory of the same period. It is to this colonization of the home as a leisure workshop that we now turn.

Home workshops were the rule in preindustrial times, almost by definition: industrialization moved production out of the home and into factories, notoriously, places where no one lives.[131] Hedonization reversed this trend, bringing artisanship back into spaces controlled by artisans. Carpentry and woodworking were occasionally mentioned in the kinds of early modern handbooks for gentlemen that we examined in Chapter 3 and had vigorous professional literatures by 1800, but indoor manual arts for men, except for brewing, seem to have hedonized later than those associated with women, like needlework and cooking. [132]

As historian of hobbies Steven Gelber has noted, woodworking of all kinds, and later, some types of metalwork, began to hedonize noticeably in the nineteenth century.[133] Their literatures had begun to diverge from those of professional carpentry, fretsawing, turnery, and joinery by 1843; and amateur turners were sending excited descriptions of home workshops to *Scientific American* by 1875.[134] *Popular Science Monthly* published home-workshop articles beginning in 1872, and the monthly magazine *Wood-Worker* began publication in Indianapolis in 1882. According to WorldCat by 1910 an average of five titles a year were appearing on woodworking, ten on carpentry, two on turning, and one or two on joinery. Like plain sewing, woodworking acquired a school workshop literature at the turn of the twentieth century.

The highly gendered stereotype of the "tool guy," although not yet known by this late-twentieth-century term, was recognizable in both Britain and the United States by 1875 and continued to gain articulated definition in the twentieth century.[135] In the late 1940s and early 1950s in America, returning World War II veterans plunged into the joys of home ownership and domestic artisanship in record numbers, and the number of titles published in woodworking, carpentry, turnery, and joinery began to rise steadily in every decade through the 1990s. In the decade 1931–40, 203 titles were published, and 328 appeared between 1941 and 1950. By the decade of 1971–80, an average of about a hundred titles a year were being published on wood artisanship, after which the rate of increase leveled off. For the decade ending in 2000, 1,246 titles are documented in World-Cat. The consumers of this flood of information were, for the most part, the kind of family man described in Jeffrey Eumenides' 1994 novel *The Virgin Suicides*. The adolescent male narrator, somewhat dazed by the heady presence of young women at a family party, is distracted by their father, clearly a serious tool guy: "Mr. Lisbon opened his tool kit. He showed us his ratchets, spinning them in his hand so that they whirred, and a long sharp tube he called his router, and another covered with putty he called his scraper, and one more with a pronged end he said was his gouger. His voice was hushed as he spoke about these implements, but he never looked at us, only at the tools themselves, running his fingers over their lengths or testing their sharpness with the tender bulb of his thumb. A single vertical crease deepened in his forehead, and in the middle of his dry face his lips grew moist."[136] So familiar was this iconic tool-cherishing male figure by then that a television series based on a caricature of him, *Home Improvement*, drew laughs from 1991 to 1999.

As we have seen, one of the historical market niches of tool guy culture in the United States was modelmaking, especially railroads, the example Steven Gelber uses as the cover illustration for *Hobbies*. Joshua Lionel Cowen began making toy trains in 1900 and by 1953 had achieved sales of $33 million, a development paralleled by the model railroad literature. In the first decade of the twentieth century, only six titles were published, and nine in the following decade. Between 1921 and 1930, however, the number of new titles more than doubled, to twenty-six. During the Depression, seventy-one new titles on model railroads were published, and eleven serial titles began publication, including *Model Railroader* in 1934. While the rate of publication on this subject appears to level off for two decades between 1971 and 1990 at a little over three hundred titles per decade, the number rose again sharply in the nineties, to 422 titles, including eleven new serial publications. Sam Posey's 2004 *Playing with Trains* may shed some light on this pattern; born during World War II, Posey returned in retirement to what had been a boyhood enthusiasm, a pattern probably now replicating itself among other retirees.[137] Posey eloquently describes the pleasures of the model railroad: the sense of power and control, the challenges of verisimilitude—he was frustrated by the unrealistic third track in his boyhood Lionel layout—and the ability to manipulate a system without any real world consequences, creating miniature train wrecks in which only plastic cows suffered. For adult men, the camouflage of virtue associated with model trains was that this hobby was an activity in which they could, ideally and traditionally, bond with their sons.

Ship models, of course, had their origin in work but had begun to enter the hobby market in the late nineteenth century. Collectors and museums were by then already at work cataloging and exhibiting the boat-shaped decorative metal "nefs" and shipbuilding models of earlier days, and manuals for building toy boats as a juvenile recreation had begun to appear by the 1860s.[138] Between 1881 and 1890, ten titles on model boatbuilding were published, and the number of titles remained low until the 1920s, when sixty-seven titles appeared, including two new serials. By the 1970s, the number of titles had reached a peak of 241 titles, five of them periodicals.

Counterintuitively, the making of model airplanes as a recreational activity appears to have preceded the making of model automobiles, perhaps because airplanes were more glamorous than cars, providing a relatively safe outlet for boyhood fantasies of flight.[139] A dozen titles on building model aircraft were published in the decade 1901–1910, and none at all on model cars, which did not

Model-building, typically a "guy" hobby, can also include the competitive sport of model racing. These Frenchman in 1946, enjoying some recreation after six years of war, prepare for serious competition. Modern steam-tractor rallies exhibit the same atmosphere of hedonized competitive intensity. Photo © Jupiter Images, reprinted by permission.

acquire a significant literature until after World War II. Model aircraft titles increased to 151, including eight serials, in the 1940s before dropping and leveling off during the two succeeding decades to a little over 80 titles each decade. In 1971, the subject apparently returned to fashion, for 196 titles were published in that decade, 9 of them periodicals; in the 1990s, 231 titles were published, of which eight were new serial titles. By that time, the literature of model autos as a hobby was not far behind, at 177 titles, including ten serials.

Leathercraft, a much less popular pastime than the various crafts associated with wood, but one which, like carpentry, requires for proper enjoyment a full and effective set of tools (though less workspace), had a tiny literature before 1900; for this period fewer than twenty titles are documented in WorldCat, most of them by or about professional tanners and leatherworkers. By the first decade of the twentieth century, however, leather had achieved a modest beachhead in the world of hobby artisanship, due in part to the influence of the Arts and Crafts movement, with sixteen titles published between 1901 and 1910.[140] In the following decades, the output of book titles more than doubled, and histories of the art began to appear, along with the inevitable connoisseurs' guides.[141] In the 1920s, hobby titles began to appear linking leatherwork both with embroidery, with which it had traditionally been associated in high-end professional work such as glovemaking and saddlery, and with the skill set of the scouting movement, which, in the United States, explicitly drew on Native American craft traditions.[142] By the 1970s, when the hobby reached its publishing peak of 195 titles, home artisans in leather had extended the craft even into the messy and odiferous art of home tanning.[143]

Ceramics (or "pottery," in the Library of Congress subject terminology) is one of the most space- and energy-intensive of the home-workshop hobby crafts, with its requirement for a kiln capable of heating earthenware to 800°C and porcelain to 1400°C. The making of pottery has a rich professional literature, in more languages than most of us can read, that has, since the eighteenth century, been thoroughly admixed with the complexities of chemistry and physics and their associated mathematics.[144] Histories and works on collecting ceramics began to appear by 1800 and were numerous by the 1870s.[145] There were, however, few amateurs' manuals before the twentieth century, except those for china painting, which was a feminized hobby craft in Britain by 1877 and in the United States by 1899. The success of china painting in the New World was partly the result of active promotion by American ceramicist Mary Louise McLaughlin and Arts and Crafts figure Adelaide Alsop Robineau. Both of these women found in china painting a way around the traditional exclusion of women from the potter's wheel and kiln; they did not fire the painted-on designs themselves.[146]

Despite the capital investment and knowledge demands of ceramics as a hobby, it became popular in the twentieth century for both men and women, with groups of hobbyists frequently sharing the expense of a kiln at a central location, such as a craft center, shop, or school. For the 1880s, WorldCat lists 198 publications on pottery, including three serials beginning publication in that decade, but by the period 1901–1910, these numbers were up to 462 titles total, with thirteen serials. Publication slowed slightly in the 1920s and then took off. By 1941, manuals for amateurs represented a small but visible component of the annual publication output on ceramics.[147] Some of these titles addressed pottery as a juvenile recreation, and several were manuals for use in school that taught handicraft classes.[148] One student of hedonizing technologies even anticipated the theme of the present work by writing a 1938 thesis entitled "The Development of the Pleasure Crafters Motor Driven Pottery Wheel."[149] Between 1961 and 1970, the number of ceramics titles nearly doubled over the previous decade, from 861 to 1,421, as baby boomers discovered the joys of the potter's wheel. More than 2,700 titles on ceramics and pottery were published in the 1990s, including a record thirty-eight serials, up from 861 titles in the period 1951–60.

Hedonizing the Outdoors

As we observed in the first two chapters, some outdoor activities, such as gardening and hunting, were hedonized early by wealthy elites, and these forms of leisure, along with fishing and boating, acquired literatures of pleasure in the early modern period. Camping, driving (first coaching and then motoring), shooting, and riding were all hedonized in the nineteenth and twentieth centuries from roots in earlier periods. Camping, for example, was historically counted among the pleasures of hunting, with its necessary corollary, outdoor cooking.[150] As an elite activity, however, hunting from camps involved the employment of bearers, cooks, laundry help, and other auxiliary personnel who probably did not especially enjoy the experience. As these activities democratized, the paths of hunting, fishing, and camping partially diverged, so that while combining these activities is still popular, camping became, in the nineteenth century, an activity that could be pursued by persons of moderate means, with no other objective than to go to some suitable location, set up camp, and enjoy the inconvenience of living outdoors. In the twentieth century, camping itself became subdivided into a great number of leisure options, which by the 1950s included a trend that eventually required the nouns-to-acronym-to-verb neologism "RVing."[151] Boating and fishing underwent similar transformations as leisure activities, with some elements of both agonizing back into work, as hobby fisher-folk and boaters turned their leisure and/or retirements into careers. Profes-sional competitive bass fishing and the operation of pleasure boat excursion en-terprises like Windjammer Barefoot Cruises are two of many examples of this backflow into labor.

The early hedonization of gardening also produced a copious literature as soon as printing was available as a means of communication, and WorldCat doc-

For some hobby gardeners, the impracticality of the product is part of the attraction, as in the case of amateur growers of giant vegetables. This photo of "Dr. Mathews holding giant cabbage" was taken c. 1903. © Corbis, reprinted by permission.

uments the large scope of gardening publications for both professional and amateur, with more than twelve hundred titles published before 1800. Even though
the total number of gardening titles published between 1800 and 1890 was only
in the hundreds per decade, dozens of serials began publication in every decade
of the nineteenth century. In the first ten years of the twentieth century, 860 gardening titles were published, of which nearly 70 were new serial publications. In
the following decade, the number of titles increased into four figures, and in
1924, the assertively hedonizing *Better Homes & Gardens* began its long career as
a flagship magazine of amateur horticulture and of home-ownership-as-hobby.
In the 1990s, more than 8,800 gardening titles were published, and 243 serial titles began publication. For the new serials, this was not even a record: 275 had
started up in the 1980s.

Gardening retailing also grew rapidly after World War II. As can be seen in
Table 4.1, the number of lawn and garden supply retailers ballooned from 2,700
in 1958 to over 30,000 in 1997. (Before 1958 the category was aggregated with
"Feed, Farm, Garden Supply Stores," but lawn and garden supply soon lost its association with rural life.)

Hunting, too, as one might expect, got an early start in the guides and handbooks literature, as we observed in Chapter 2, with 534 works on the subject appearing before 1800. In the nineteenth century, hunting for the sheer enjoyment
of killing, competitive (usually masculine) achievement, and the related display
of trophies was regarded as the most gentlemanly form of the sport, as opposed
to "mere pot-hunting," which was beneath the dignity of aristocrats like the Russian Grand Duke Alexis, who hunted buffalo with William Cody in the western

TABLE 4.1.
Lawn and garden supply retailing in the United States, 1958–1997

	No. of retail establishments	Retail establishments with payroll	Sales, all establishments (in $1,000s)	Sales, establishments with payroll (in $1,000s)
1958	2,735	1,783	191,039	176,999
1963	3,518	2,756	279,296	268,624
1967	3,031	2,754	399,118	380,208
1972	8,131	3,849	829,547	694,807
1977	15,487	6,921	1,788,737	1,584,489
1987	20,715	10,692	5,808,894	5,410,774
1992	22,062	10,857	6,772,668	6,327,846
1997	30,066	21,201	32,114,031	31,677,905

Source: Data from *Census of Retail Business* (before 1972) and the *Census of Retail Trade* (1972 and later) for
Standard Industrial Classification (SIC) 526 / North American Industrial Classification (NAIC) 4442.
  Note: Here and in Tables 4.2, 5.1, and 5.2, the year 1982 is omitted because no data were available for
that year.

United States in 1871–72.[152] Another titled hunter, the duke of Saxe-Coburg, reportedly had, in 1894, "over seven thousand horns and antlers" of chamois alone mounted on the walls of his Tyrolean hunting seat, Hinter Riss.[153]

Like gardening, hunting gained popularity steadily as a book topic during the nineteenth century, with 114 titles published between 1821 and 1830, and 560 between 1871 and 1880; but unlike horticulture, the number of new serial titles was relatively small, with only eleven in the earlier decade and twenty-one in the later one. The 1930s saw only about half as many new serials on hunting as on gardening, but in the 1960s, hunting serials caught up with and then surpassed gardening in numbers of titles published per decade, with 508 beginning publication in the 1990s, about twice as many as on gardening. As in the cases of candymaking and cake decorating, this difference in timing and technical paths is difficult to explain.

The pre-1800 literature of fishing—of which 490 titles were documented in WorldCat at the time of my search—is rich in references to pleasure, contemplation, and the leisure artisan's love of tools and gear, all as familiar to anglers as to needleworkers. In the rhetoric of pleasure, Isaak Walton (1593–1683) set the tone for future centuries with his gentlemanly enthusiasm for the sport; his title, *The Compleat Angler,* must be one of the most imitated in English nonfiction.[154] James March continued the hedonizing rhetorical tradition in his 1833 *Jolly Angler,* a title that was to enjoy a second immortality as the name of the Manchester pub in which the Beatles performed in the 1960s. Like knitting, fishing has literatures of the activity both as a path to spiritual wisdom and as a road to perdition, some of the latter written by wives of fishermen.[155] While fishing was historically a largely, but not exclusively, male recreational activity, works began to appear in the twentieth century encouraging girls and women to participate; and both Girl and Boy Scouts were taught the parallel arts of fishing, fly tying, and rod-making, as well as how to cook their catch.

By contrast, women and girls represent the principal market for the American hedonization of the horse. In 1910, there were almost 20 million horses and ponies on U.S. farms; in 2000 about 5 million American households among them owned about 7 million horses, although as a means of transportation horses have been obsolescent for the better part of a century. Of these equines, 5.1 million were in 2001 owned as pets, up from 4 million in 1996. Of households owning horses as pets, in 74.9 percent of them the animals were cared for by women and girls in 2001.[156]

For some police officers and ranch hands, riding horses is a component of labor; Amish farmers farm and travel successfully with horse power; and busi-

Some anglers rarely fish but spend hours engaged in the craft of tying flies: A fly-tier in Monmouthshire, Wales, c. 1950. Photo by Kurt Hutton © Corbis, reprinted by permission.

nesses like luxury cab services and riding stables make economic use of horses, mules, and ponies as a means of getting to and from work, or of performing it. But aside from these few exceptions equines in the American social landscape are almost entirely a hedonized phenomenon.[157] In 1999, the U.S. horse industry produced $25.3 billion in goods and services, employing 619,400 persons.[158] Racing, which accounts for 25 percent of this total, is at least nominally a business, of course, but few racing-stable owners would argue for Thoroughbred horse ownership as a means of making a comfortable living. Most of the business uses of horses are basically about pleasure: the owners and operators of riding stables, dude ranches, and horse-drawn vehicle services may be working hard, but their customers are out to have fun. There are exceptions, like the use of horses by mounted police, for herding cattle, and for search and rescue in difficult terrain, but for the most part, we Americans (and for that matter, Britons and Europeans as well) keep horses in our lives for the same reason we have dogs and cats: because we like them and enjoy their company.

The hedonization of the horse began in the mid-nineteenth century, when railways supplanted the traditional stagecoach as a means of getting from one place to another. Stagecoaches had a deservedly evil reputation, especially before the invention of the leaf spring at the turn of the nineteenth century: strap-hung coaches induced nausea; passengers were crammed together in a claustrophobic and usually odiferous space; the vehicles were inherently unstable, especially when carrying heavy luggage on the roof, and were liable to bog down in mud on the then-execrable roads. Interiors were described as dirty, shabby, and harboring vermin, especially fleas; schedules were unreliable; drivers were frequently rude, dishonest, violent, drunk, or all four; and the inns and taverns required both by the need to victual and lodge passengers and by the staging system (changing and relay of horses) had, if anything, worse reputations than the coach services.[159]

Nevertheless, no sooner did the stagecoach begin to disappear, first from the British countryside and then, somewhat later, from the American, than affluent young men—significantly, those too young to remember the miseries of riding in actual commercially operated stagecoaches—seized on the technology and converted an unsatisfactory means of transportation into the recreational activities of coaching and driving.[160] William Thackeray was astonished and amused by this development, remarking in 1854 on the notorious crudeness of early-nineteenth-century coachmen and their later emulation by young male gentry: "To give and take a black eye was not unusual nor derogatory in a gentleman; to drive a stage coach the enjoyment, the emulation of generous youth. Is there any young fellow of the present time who aspires to take the place of a stoker?"[161] He would no doubt be equally amazed that the answer to his question is now a resounding "yes," as steam power, now obsolete for rail transportation, has passed into nostalgia and hedonization in the same way that stagecoaches did a century earlier. The leisure activities of coaching and driving acquired a literature of pleasure in Britain and the United States late in the nineteenth century, as did steam transportation, interestingly, at about the same time, as hobby and juvenile modelmakers, aspiring not to Thackeray's stoking but to engineering, tinkered with steam technology in home workshops.[162]

In the second half of the nineteenth century, a number of aspects of life in the American West, including its horse culture, were systematically glamorized and exploited commercially, with spectacular success, most notably by William "Buffalo Bill" Cody and his associates, including a woman, Annie Oakley, who, like Cody, had been a professional hunter of game meats for food service.[163] Part of this process of glamorization was the making of traditional cattle-ranch *technē* into a sport, the modern rodeo, in which persons, with what seem to some of us inexplicable tastes in recreation, ride bulls, broncos, and even bison for the competitive pleasure of being thrown from them.[164] The rodeo events of bull riding and bull dogging—leaping onto a running calf from horseback and throwing it to the ground by the horns—have no practical or historical analogues in cowboy working life.[165] Other events, such as racing quarter horses around barrels, are perhaps somewhat less gratuitously hazardous and are more obviously connected with historical requirements for agility and changes of direction in pursuit of cattle. As Jane Austen astutely observed, "One half the world cannot understand the pleasures of the other."[166]

Similarly, one might intuitively suppose that intentional destruction and violence would resist hedonization, but just as children historically have played at being deadly and dangerous adults, so do adults adapt military technologies and skills to less fatal purposes than their battlefield counterparts. We have already noted this tendency to glamorize the robust interactions of soldiers, cowboys, and hunters with nature, in the discussion on hedonizing the outdoors and on outdoor cookery; we now turn our attention to the paths taken by warfare into the world of hobbies and leisure.

Leaving aside, as I have throughout, the issue of children's toys (e.g., toy soldiers) as miniature versions of adult technologies for the perhaps more complex adaptation of battlefield technologies to juvenile recreation (e.g., real bows with blunt-tipped arrows), the earliest trend I can readily identify as a hedonized paramilitary activity is that of scouting, the brainchild of British aristocrat and lieutenant-general Robert Baden-Powell (1857–1941), who believed that boys as well as men benefited from exposure to military virtues and the outdoor life. Testing his concept with a small troop in 1906–7, he built the foundations of Boy Scouting in Britain, and wrote a training manual, *Scouting for Boys*, which went into its thirtieth edition in 1957 and was still selling fifty thousand copies annually in the early 1960s.[167] Confronted at London's Crystal Palace in 1909 by a group of girls indignant at their exclusion from scouting, Baden-Powell established the Girl Guides the following year; at this writing (fall 2006), there are about ten million Guides worldwide. Both organizations remain proudly and defiantly gender-segregated.[168] For both boys and girls, the paramilitary uniform, the system of ranks and honors, outdoor life, and manual crafts were and are central to the scouting scheme.

Baden-Powell's vision for British boys was interpreted for Americans by wildlife author Ernest Thompson Seton (1860–1946), who introduced a major infusion of Native American lore and crafts, including "Indian signs and blazes," the making of tepees, and a number of Indian games In the 1910 edition of his *Boy Scouts of America: A Handbook*, he and other contributors presented a list of

Scouting is a hedonized paramilitary activity, mimicking not only military training and accomplishments, but also music, uniforms, and medals. Poster for the Boy Scouts' Rally in 1909 at the Crystal Palace in London, during founder Baden-Powell's lifetime.

"honors" to be earned for many of the hedonized activities discussed in this book: riding, hunting, shooting, "campercraft," fishing, and photography.[169]

The Camp Fire Girls, which describes itself as "the first nonsectarian, interracial organization for girls in the United States," was established in 1910, and now, as Camp Fire USA, has more than 700,000 members of both sexes. Juliette Gordon "Daisy" Lowe (1860–1927), after a 1911 meeting with Baden-Powell in Britain, came home to Savannah, Georgia, to organize the first U.S. chapter of what were then the Girl Guides (later the Girl Scouts) in 1912; the Girl Scouts of America now has 3.7 million members.[170] While there is a strong element of community service in modern scouting for both girls and boys, the activity retains the focus of its turn-of-the-century founders on healthy, enjoyable, and productive outdoor recreation. Of these organizations, only Camp Fire has dispensed with Baden-Powell's characteristic paramilitary uniforms and with the system of ranks, promotions, and service decorations, along with gender segregation.

Reenactment of battles by modern hobbyists in historical dress, made famous nationally by Ken Burns' 1989 *Civil War* documentary, is a leisure activity with antecedents in noble pageantry of the late medieval and early modern period, but the serious, research-intensive pursuit of military and historical verisimilitude is, according to *Wikipedia*, "an invention of the twentieth century."[171] Although most Americans associate reenactment with the Revolutionary and Civil Wars, European and British reenactors interpret their history with impersonations of ancient Roman, Napoleonic, and even World War I and II military characters.[172]

The American phenomenon of reenactment seems to have roots in two traditions: the reenactment of famous battles as a component of military training and the therapeutic reliving of battles by veterans of them, an irony of which I shall have more to say later on. Historian Jenny Thompson, writing of reenactment in 2004, says: "As early as 1822, twenty American Revolutionary War veterans engaged in a reenactment of their 1775 clash with British troops at Lexington; in 1902, Crow Indians and state militiamen reenacted 'Custer's Last Stand' near Sheridan, Wyoming."[173] There is an evocative account of the fiftieth anniversary reunion of Civil War veterans at Gettysburg in 1913 in which the activity is clearly intended as therapeutic:

> The climax of the gathering was a reenactment of Pickett's Charge. Thousands of spectators gathered to watch as the Union veterans took their positions on Cemetery Ridge, and waited as their old adversaries emerged from the woods on Seminary Ridge and started forward toward them again, across the long, flat fields. "We could see," [Philip] Myers [b. 1895] wrote, "not rifles and bayonets but canes and crutches. We soon could distinguish the more agile ones aiding those less able to maintain their places in the ranks."
>
> As they neared the northern line, they broke into one final, defiant rebel yell. At the sound, "after half a century of silence, a moan, a sigh, a gigantic gasp of unbelief'" rose from the Union men on Cemetery Ridge. "It was then," Myers wrote, "that the Yankees, unable to restrain themselves longer, burst from behind the stone wall and flung themselves upon their former enemies . . . not in mortal combat, but reunited in brotherly love and affection."[174]

It has been observed that few modern reenactors are combat veterans, so healing from the psychological wounds of warfare cannot be their objective, as it evidently was for the men weeping and hugging each other in front of the wall at Gettysburg in 1913. The modern hobby seems to have originated with the centennial of the Civil War, when the First Battle of Manassas/Bull Run was reenacted in July 1961. Since then, all types of battle reenactment have attracted tens of thousands of men, and a much smaller group of women, to the joys of camping out, wearing period clothing, and pretending to kill each other. The small number of women— about 3 percent—impersonate camp followers, wives, sutlers, laundresses, and even female soldiers passing as men, a choice of role that generated a lawsuit in which a female reenactor had to demonstrate her right to participate in Civil War battle reenactment with an array of historical documentation of women serving as men in the original conflict.[175] For women with less interest in combat, needlework and outdoor cookery often play significant roles in reenactment.[176]

Paintball is still too young a sport to have made much of an impression on scholarly literature, but it has evolved, since its origins in the 1980s, from a tracking and woodcraft pursuit like some forms of scouting, involving the stealthy capture of flags with or without recreational gunplay, to a paramilitary activity that now more closely resembles battle reenactment than anything Ernest Thompson Seton would have endorsed for American boys.[177] Seton's 1910 *Handbook* does, however, describe a game attributed to Baden-Powell that reads like distant ancestor of the sport:

> A lion is represented by one scout, who goes out with tracking irons on his feet . . . and six lawn-tennis balls or rag balls. He is allowed half an hour's start, and then the patrol go after him . . . each armed with one tennis ball with which to shoot him when they find him . . .
>
> When they come near his lair the lion fires at them with his tennis balls, and the moment a hunter is hit he must fall out dead and cannot throw his tennis ball. If the lion gets hit by a hunter's tennis ball he is wounded, and if he gets wounded three times he is killed.[178]

Modern paintball warriors, of course, use much more sophisticated technology. According to James Gibson, author of the 1994 *Warrior Dreams: Paramilitary Culture in Post-Vietnam America*, the technology originated with a device called Nel-Spot, made by Daisy of air-rifle fame, "intended for use by farmers and ranchers to mark trees and animals with watercolor paint contained in gelatin capsules." Once paintball battle parks began to appear in the 1980s—Gibson reports a thousand in the United States by 1988—tinkerers and manufacturers began adapting the air-gun equipment to faster rates of fire and developing refinements like paintball land mines and tanks. Gibson notes that only a tiny minority of paintball soldiers have ever been in any kind of real live-fire combat,

Paintball, which became popular in the United States after 1980, has origins in battle reenactment and scouting. These paintballers and the paintball tank are participating in the Battle for Serenity Valley, Wisconsin, in July 2008. Photo © Kristina Jager, reprinted by permission.

a finding consistent with some observations of the demographics of historical battle reenactment. Real combat soldiers have been there, done that, and apparently are not inclined to construe mock warfare as recreation.[179]

Hedonized Communication

Most historians address the history of communication as a functional issue—telegraphy, for example, as a means of getting a signal from one place to another. A minority, including Kristen Haring writing on ham radio in 2003 and Claude Fischer on telephony in the late 1980s, have observed and described the pleasure that communications-technology users have in constructing, configuring, and using the devices to communicate as a leisure activity.[180] Of course, the activities of Internet bloggers and cell-phone users—many of whom alleviate the boredom of, for example, waiting in line by calling their friends for a chat—are obvious cases of communication as a pastime. I intend, however, to steadfastly ignore these intellectual temptations in order to concentrate on the role of artisanship and skill in leisure technologies of communication. Of many possible examples,

I have chosen ham radio, high fidelity, and photography. All three are examples of what are called "scientific" hobbies, in that serious practitioners usually need to acquire a fair degree of technical expertise in the physical sciences related to their craft; all three are in the tradition of affluent Western mostly male hobby experimentation that found expression in earlier centuries in amateur involvement with scientific societies. In all three cases, the technology was hedonized as rapidly as it developed, with the production and hobby paths diverging early but remaining in close touch with each other as technical ideas and innovations crossed hobby-professional boundaries in both directions. Men who had built amateur radio stations as boys, for example, often became radio industry engineers; and hobby photographers sometimes decided to take up their art as a profession.

Historian of technology Kristen Haring has vividly evoked the gendering of ham radio as a hobby, with its traditional designation of wives as "XYLs," Ex–Young Ladies, suggesting that women represented a kind of adversarial presence in the hobbyists' households. Keir Keightley has shown the same type of gender tensions for audiophiles in his evocatively titled 1996 *Popular Music* article "'Turn It Down!' She Shrieked."[181] Both hobbies are historically and robustly male, and in the United States, white and middle-class and up, in part because of the expense of components and information. Boys not only enjoyed being in control of the high technologies of the day but could dream of participating in its larger world—and indeed, many made good economic use of the skills they learned.[182] Amateur radio operators of the early twentieth century might even be able to report a sinking ship and save lives. Historian of technology Susan Douglas summarizes the manly pleasures of early-twentieth-century ham radio, in her 1987 *Inventing American Broadcasting, 1899–1922,* as follows:

> The boy would enter a new realm in which science and romance commingled. He became an explorer. He both triumphed over and was in harmony with nature. Through wireless, the experimenter went through the looking glass, to a never-never land in which he heard the disembodied "voices" of ships' captains, newspaper men, famous inventors, or lovers passing in the night . . . He could please his parents by acquiring this instructive hobby, and he could defy them by using it, without fear of being discovered, to misbehave. In this realm, in the "folds of the night," by mastering a new technology while letting his imagination and his antisocial inclinations run loose, he could be, simultaneously, a boy and a man, a child and an adult.[183]

Boys formed amateur radio organizations in the United States as early as 1909, and they remain the principal consumers and practitioners of this hobby technology despite strenuous efforts since the 1980s to interest girls and women.[184]

Ham radio is represented in library catalogs by two Library of Congress sub-ject headings: "Radio—Amateurs' manuals" and "Amateur Radio Stations." The first results for either heading in WorldCat appear in the 1920s, when fifteen ti-tles identified as amateurs' manuals were published, including three serials, and one book on amateur radio stations. This suggests that the hobby was already well under way by this time and that, as is often the case, library catalogers were struggling to catch up with a moving bibliographic target. For the Depression era, we see only five manuals and seven radio-station titles, with only the latter listing a new periodical title. In the 1960s, however, 135 manuals and 81 titles on ham radio stations were published, including three new serials in the former category and eight in the latter.

The path of high fidelity as a hobby shows a similar historical pattern. Edward Tatnall Canby, writing of high fidelity in 1955, expressed the pleasure audiophiles experience in their hobby in technologically enthusiastic terms: "As a generic term, hi-fi does not 'mean' but 'is.' First of all it is a hobby area. A business, a con-glomeration of manufactured equipment, a vast array of custom-designed cabi-nets, tubes, dials, tangles of wire and pounds of solder, not to mention miles of tape and shelves full of records. More than all this, hi-fi is a Way of Thinking, a means of emotional catharsis and of creative action. Hi-fi is a big-time hobby, backed by a serious art, and for a big-time hobby we need big-time coverage. High fidelity! A good Americanism."[185]

While, in theory, high fidelity is about the "purity" of recording and playback of music—what Eric Rothenbuhler and John Peters, writing of the "analog renais-sance" in 1997, call the "moment of fidelity when the equipment disappears and the pure music flows through"—it has frequently been noted that audiophiles pursue a goal beyond music, what audiophile Norman Eisenberg called in 1959 "sound for sound's sake."[186] There are, as historian of technology Joseph O'Con-nell has observed, elements of snobbery and connoisseurship associated with the cultivation of a "golden ear," the ability to distinguish high-quality sound repro-duction from its lesser counterparts. Like embroiderers, the practitioners of this art are prone to collecting artifacts and information about it and to see more virtue in past expressions of the craft than in those of the present.[187] Digital audio, for example, is deeply mistrusted by audiophiles, and some even eschew solid-state technology in favor of tubes. As we have seen, it shares this conservatism about tools and equipment with many other artisan hobbies and leisure activities.

Like the literature of ham radio, that of audiophilia was born in the 1920s, with a meager two titles, one of which was the serial *Gramophone*, which began publication in 1923. Like that of their radio counterparts, the expansion of published titles in the subject areas "Stereophonic Sound Equipment" and "High-Fidelity Sound Systems" occurred in the 1950s and 1960s, with more than 150 titles, including more than 30 new serials, in each of these decades. This literature, however, apparently reached its peak in the decade 1971–80, with 344 titles, of which 48 were new serials. In the 1990s, the number had dropped to 124 titles, of which 27 were new periodicals.

Uniquely among the technologies I have discussed here, photography was apparently a smash hit among both professionals and amateurs while still in its infancy. WorldCat lists 77 titles on photography, among them three serials, for the period 1831–40; in the following decade 186 titles appeared, including ten serials. Between 1851 and 1860, 509 titles were published, of which a whopping 73 were serials. Photographic works continued to appear at this very high rate through the nineteenth century and for the first few decades of the twentieth before exploding into 1,198 titles in the 1930s, with 139 serial titles beginning publication during that time. In the 1990s, 7,389 titles appeared, of which 241 were serials. Many of these thousands of works were bought and read by both amateurs and professionals; in some cases the titles make it clear which audience is being addressed, but in most, the readership is assumed to be any and all photographers.[188]

The growth in publishing was paralleled, unsurprisingly, by rapid increases in retail sales in the United States (see Table 4.2). In 1929, the Census of Business counted 710 camera and photographic stores selling about $28 million in product; ten years later, despite the intervention of the Depression, there were 1,112, and 3,030 by 1948, with total sales of $202 million.

Even in the mid-nineteenth century, the enthusiasm of amateurs for photography was front and center in the field, helping to drive economic and technical growth and development.[189] By 1940, Robert Johnston could plausibly assert in *American Speech* that photography was "one of the most popular as well as one of the most expensive hobbies in America."[190] Like other hobbies we have discussed here, cameras, gear, and photographs quickly became "collectible" and began to appear in museums.[191] What is it about photography that made it immediately popular as a hobby? While I will leave it to the historians of photography to tease out all the elements that contributed to this phenomenon, I will suggest a few contributing factors.

First, photography is by nature a mechanical and industrializing art. Its relatively late appearance on the scene (compared with, for example, hunting and embroidery) gave it almost immediate advantages for democratization, especially after the invention of roll film and readily portable cameras.[192] Second, it ap-

Photography was hedonized almost as soon as it was invented, possibly because it appealed to so many elements of the leisure experience, two of which are highlighted in this 1916 advertisement from *American Photography:* the ability to share one's hobby accomplishments with others, and to do so in technologically sophisticated, professional style

peared during an era, the mid-nineteenth century, when science was actively pursued as a hobby by those who could afford it; photography offered, as it still does, opportunities to dabble joyously in physics, optics, chemistry, and even meteorology and astronomy.[193] Indeed, I would suggest that this affinity for combination with other leisure activities has contributed to its success; many hobbyists from the nineteenth century forward have used cameras to record and share experiences with other enthusiasts. Pictures of fishing, hunting, antique-collecting triumphs, needlework, woodworking, model-building, cake decorating, cooking, and dozens of other hobbies have allowed artisans to communicate their skills

TABLE 4.2.
Camera and photographic supply retailing in the United States, 1929–1997

	No. of retail establishments	Retail establishments with payroll	Sales, all establishments (in $1,000s)	Sales, establishments with payroll (in $1,000s)
1929	710	—	28,810	—
1939	1,112	—	32,343	—
1948	3,030	—	202,099	—
1954	2,896	2,270	278,101	—
1958	3,491	2,839	381,938	364,144
1963	3,308	2,800	378,947	365,935
1967	2,767	2,540	462,298	439,837
1972	4,743	3,271	766,737	735,640
1977	5,672	3,550	1,170,767	1,110,758
1987	5,546	3,791	2,382,750	2,294,000
1992	4,871	3,012	2,303,596	2,207,491
1997	4,507	2,843	2,348,155	2,251,690

Sources: Data from the *Census of Business* (before 1972) and the *Census of Retail Trade* (1972 and later) for SIC 5946 / NAIC 443130.

and enthusiasms across space, time, and language barriers.[194] This effect of photography has driven the production of what are called "coffee-table books" for hobbyists and connoisseurs, which are composed almost entirely of illustrations of exceptionally interesting examples of their art.

Cameras and their related gear have also appealed, understandably, to hobby tinkerers, in the same way that ham radio, classic automobile restoration, and model railroads do. In the late nineteenth century, according to *Scientific American*, it was not unusual for hobby photographers to build their own cameras and to design improvements for them.[195] Some amateurs also enjoyed the "magical" qualities of photography, making images appear and disappear by bleaching followed by an application of hypo, and "table top photography," in which powdered alum can be made to look like a landscape at sunrise, and lumps of wax stand in for icebergs.[196] Barbara Kingsolver describes a hobbyist of this latter type in her 1991 novel *Animal Dreams:* "Doc Homer made pictures. Specifically, he made photographs of things that didn't look like what they actually were. He had hundreds: clouds that looked like animals, landscapes that looked like clouds."[197]

Not only did amateur photographers have exciting options about what they did with the practice of their craft; they could also join clubs and share ideas, compete, and tinker in the company of like-minded hobbyists.[198] In this and some of its other characteristics, photography shares qualities with high fidelity and ham radio, and like them, it evolved a lively retail industry and periodical press.

Counterintuitive as it might seem, embroidery, hunting, home preserving, paintball, photography, camping, and many other hobby and leisure activities have elements in common that set them apart from technologies of market production, and evolutionary histories in which efficiency and productivity are minor concerns. What is the motivation for this broad-ranging urge to make tools into toys for grown-ups? It is to this question that we now turn.

CHAPTER FIVE

Why, When, and How
Do Technologies Hedonize?

The end of World War II brought a wave of prosperity to the United States, at the time the only major economic power in the world with an intact manufacturing base. Both men and women sought occupations for their leisure, which had been slowly but surely increasing as labor movements "brought us the weekend," as the bumper sticker has it, and shorter working days. Even non-union white-collar workers benefited from the long uptrend in nonwork hours—the "eight hours for what we will" that was the objective of labor activists on both sides of the Atlantic at the turn of the twentieth century.[1] In the case of women, part of the postwar expansion of their unpaid time was, of course, an artifact of demobilization, as women lost their wartime jobs to returning veterans.

Whether willingly or not, women returned to their homes, where the average number of hours required to maintain a home and family remained high, but women were at least no longer working a double day.[2] Men who could afford it built workshops, tinkered with their cars, built models, barbecued, took on home improvement projects, and supplied themselves liberally with both hand and power tools. Over the next fifty years, the leisure workspace of American homes expanded into rooms that either did not exist in the nineteenth century or were more production-oriented: the living room (one does not "live" in a parlor) and the family room, as well as into the kitchen, basement, garage, patio, and sometimes, usually unintentionally, the dining room.[3] Even attics were pressed into service for such expansive leisure activities as building model railroad layouts and organizing large sewing projects. Properties that had once been farms were and are often rich in outbuildings, including barns, stables, sheds, and chicken coops that could be colonized for workshops, studios and tool storage.[4]

The hobby market responded to these developments with predictable alacrity: hobby, toy and game shops, which previously had been folded into "All other retail stores" in the U.S. *Census of Business*, emerged as a separate heading under

TABLE 5.1.
Hobby, toy, and game retailing in the United States, 1958–1997

	No. of retail establishments	Retail establishments with payroll	Sales, all establishments (in $1,000s)	Sales, establishments with payroll (in $1,000s)
1958	4,489	2,437	193,227	163,819
1963	4,278	2,726	257,684	234,806
1967	3,213	2,838	317,237	315,876
1972	10,476	4,267	907,924	811,460
1977	18,543	6,570	1,768,314	1,554,466
1987	28,052	9,629	7,451,069	7,031,359
1992	37,747	10,860	11,297,797	10,627,271
1997	31,030	10,824	15,110,488	14,388,277

Sources: Data from the *Census of Business* (before 1972) and the *Census of Retail Trade* (1972 and later) for SIC 5945 / NAIC 451120.

SIC code 5999 in 1958.[5] In that year, enumerators counted nearly 4,500 such establishments, about half of them with payroll, with total sales of a little over $190 million (in 1958 dollars); the number of establishments was down slightly in the economic census of 1963 (see Table 5.1). Less than ten years later, however, there were more than ten thousand hobby, toy, and game stores with total sales approaching a billion dollars annually. By 1977, there were more than 18,000 establishments grossing $1.77 billion. The toy and game component of this retail channel accounts for some of this growth, of course, but other kinds of hobby retailing, such as gardening and needlework, were also in growth phases, as we shall see.

Leisure Needlework in the Postwar United States

In the U.S. Department of Commerce *Census of Business*, the retail industry in needlework and home sewing materials was not listed separately from department and other general merchandise stores until 1958. By then, it was deemed significant enough to merit its own category, showing more than 1,500 retail establishments engaged solely in the sale of supplies for needlework and sewing and about 100 more operated as leased departments in larger stores. In 1972, the census reported almost 20,000 separate establishments. In the same period, gross national sales at retail rose steeply from just over $43 million annually to more than $1.5 billion.

The 1960s and 1970s were a time of rapid growth and expansion in the American needlework community. Not only did needleworkers become more numerous and visible in the population; they also explored new techniques and materials, bringing to the medium both the technical expertise of traditional needle-work and innovative ideas about technology, media, and aesthetics. The number of needlework magazines and books published annually more than tripled be-tween 1967 and 1979. Embroiderers' guilds, leisure organizations taking their name from the much older professional tradition, had been established in Brit-ain early in the twentieth century. They were a new phenomenon in the United States in 1967, but by 1977 the Embroiderers' Guild of America numbered more than ten thousand members, with new chapters forming every month.[6] The same decade saw the establishment and growth of other U.S. needlework orga-nizations, including the National Quilting Association, the Counted Thread So-ciety of America, and the National Needlework Association. Older organizations, such as the International Old Lacers, increased their membership significantly during this period. American women's involvement in needlework grew rapidly, as did their spending in the needlework and home sewing market (see Table 5.2). In the 1970s, needlework was the second largest hobby market in the United States, with gardening in first place.[7]

Growth brought changes in the structure of the industry that followed trends in retailing generally, with multi-unit firms (chains) increasing in number from a scant 239 in 1963 to more than four thousand, or about a fifth of all needlework retailing establishments, in 1972. The same increase in the average size of needle-work retailers is evident in the shift from establishments owned by a single pro-prietor to an increasing number managed by corporations.

Contributing to and feeding on this revival of interest in needlework were trends in hobbies and leisure activities generally; as we have noted, after World War II, traditional revivalists in lithography and handweaving brought new and

TABLE 5.2.
Needlework and sewing retailing in the United States, 1958–1997

	No. of retail establishments	Retail establishments with payroll	Sales, all establishments (in $1,000s)	Sales, establishments with payroll (in $1,000s)
1958	1,564	774	43,728	34,498
1963	2,566	1,440	86,217	74,493
1967	2,193	—	172,284	—
1972	19,784	11,414	1,585,230	1,466,079
1977	19,057	10,064	1,958,001	1,798,045
1987	21,742	9,632	3,137,980	2,835,612
1992	19,612	8,264	3,857,265	3,576,322
1997	19,676	6,590	3,477,978	3,182,916

Sources: Data from the *Census of Business* (before 1972) and the *Census of Retail Trade* (1972 and later) for SIC 5949 (formerly 539) / NAIC 451130.

liberating ideas to these media. The hobby magazine *Handweaver & Craftsman* began publication in 1950. In embroidery, pioneers like Mariska Karasz wrote guides that took the art seriously, even changing the name of a long-familiar stitch, the lazy-daisy, to "open chain," an innovation celebrated decades later by fabri-crafter Robbie Fanning in the title of a new serial on the textile arts.[8] Jacqueline Enthoven and Nik Krevitsky were important contributors to this new view of embroidery as a medium for artists, of which I shall have more to say later in this chapter.[9]

As in the case of handweaving and some other hobby media, the needle arts revival of the 1960s and 1970s developed three separate but related paths. A traditional revivalist movement in crewelwork and canvas embroidery (needle-point)—led in the United States by embroiderers like Erica Wilson, Mildred Davis, Muriel Baker, Bucky King, and Elsa Williams, and in Britain by Constance Howard, Diana Springall, Mary Gostelow, Beryl Dean, and others—began to gather momentum in the 1960s and was the driving force behind the reappearance on the American scene of small, independently owned retail stores that specialized in embroidery patterns, materials, and technical assistance to hobby-ists.[10] Canvaswork engineer Sherlee Lantz produced a huge book of embroidery stitch architecture, then followed it with a book featuring the invention, for the first time in many centuries, of an entirely new canvaswork stitch, called trian-glepoint. Both works sold well despite their high prices.[11]

In this technological and aesthetic path of embroidery, much of the market was in middle adulthood and, of course, mostly female, affluent, conservative, and white, with occasional anomalies like Roosevelt Grier. While mail order and department stores competed for the needlework market overall, the high end of the needlework market could afford personal service and costly materials, and this preference is reflected in the retail growth shown in Table 5.2. Publishers like Van Nostrand Reinhold, Branford, Dover, and Scribner's in the United States and Batsford in Britain addressed this market with hundreds of lavishly illustrated titles.

At about the same time, a second strand of hobby needlework was taken up by the generation then in its teens and twenties: the hippie and peacock revolutions of flash clothing and personal adornment. Denim in particular was a favored ground material for psychedelic embroideries and patchwork in blazing colors, but any fabric from velvet to burlap could be and was rendered wondrous by needlework embellishment.[12] The styles created by this generation of needle-workers, of which the author is a member, are now regarded with awe and envy by young people who covet our decorated "real jeans from the seventies" as articles of antique clothing. The batik and tie-dyeing of our young adulthood are in revival as well. The combination of rich embroideries, laces, and patchwork with informality was an entirely new concept in dress; it was of a piece with the boomer generation's liberation of athletic footwear from the gymnasium, and of T-shirts from the stigma of underwear. Many of us in the 1970s were under the spell of romanticism and nostalgia for everything associated with our grandparent's (but not our parent's) generation, including needlework, home preserving, soapmaking, candles, and farming. As sociologist Trevor Pinch has pointed out about music, and vegetarian gastronome Anna Thomas about cooking, marijuana and other drugs had effects on the aesthetics of leisure activities that were both significant and enduring.[13]

The third path was the self-conscious product of textile artists. While Erica Wilson was designing kits that encouraged upper-middle-class women to emulate the fine crewelwork of the eighteenth century, and hippies were stitching rainbows onto their denim jackets, serious artists of both sexes, many of them university-trained professionals, undertook a full-scale exploration of the possibilities of textiles as a medium. Weavers like Jack Larsen, Anni Albers, and the Justemas made works of art for exhibitions and published their designs in glossy, heavily illustrated catalogs and manuals; the Kliots were active in reviving and redefining lacemaking and macramé; and even crochet acquired a male promoter in Mark Dittrick, author of the virile title *Hard Crochet*.[14]

In the 1970s two more trends contributed to the renaissance of needlework. One was the revival of quiltmaking and the rise of quilt collecting; the other was the feminist movement, particularly the faction of it that was concerned with the arts. Two techniques used in quiltmaking—piecing and appliqué—had been included in the embroidery revival of the 1960s, but the making of whole quilts, either as a pastime or as an activity for professional artists, did not gain traction until the 1970s.[15] Some historiographic groundwork had been laid in the 1930s and 1940s by Carrie Hall, Rose Kretsinger, and Florence Peto; and two quilter-collectors, Marguerite Ickis and Ruby McKim, had added to the store of knowledge about quilt patterns in 1959 and 1962, respectively, but the modern fashion for quilts was for the most part a phenomenon of the 1970s.[16]

While quilts had been exhibited in historical museums before 1971, the marker event for the quilt revival is generally thought to be the exhibition of quilts as abstract art at the Whitney Museum in New York City in that year, curated by Jonathan Holstein and Gail Van der Hoof.[17] Other museums quickly followed the Whitney's lead, with the collectors and connoisseurs in hot pursuit.[18] Scholars of quiltmaking began to publish, and the historical literature benefited from research by pioneers like Cuesta Benberry, Lenice Ingram Bacon, Carleton Safford, Robert Bishop, Patsy and Myron Orlofsky, and Mary Washington Clarke.[19] Artists were not slow to see the possibilities of the medium once quilts were off of beds and onto walls. By the late 1970s a number of professional artists were routinely producing and exhibiting quilts as part of their *oeuvre*.[20]

Into this, as it were, hotbed of 1970s quilt activity marched feminists like me, looking for aesthetic traditions to which we could relate our own experiences of the textile crafts. As Paula Mariedaughter recently expressed it in her undated "A Brief History of the Art Quilt,"

> In the early 70s the Women's Liberation Movement was fueled by the passion and the anger of women objecting to the exclusion of women and women's work from much of the mainstream culture including the art world. Women described taking art history courses that never mentioned one woman. Art museums and art galleries rarely exhibited art by women.
>
> Women wanted to see art by women. The pro-woman attitudes that grew from the Women's Liberation Movement applied pressure on all segments of the art world to include art by women. Women demanded that narrow definitions of what can be considered "art" be expanded to include areas where women artists predominated—like quilting and the so-called "decorative" arts.[21]

Art historian Patricia Mainardi's 1973 article, "Quilts: The Great American Art," in *The Feminist Art Journal*, later reprinted in part in *Ms.* and *Art News*, kicked off several decades of intensive research and publishing on quilts and other needlework as women's arts, a process which continues in feminist scholarship now; indeed, this book is itself a contribution to the feminist literature of needlework that saw its modern beginnings in the 1970s.[22] Artist Judy Chicago even went so far as to make of needlework and china painting an essentialist argument about a feminine aesthetic, mobilizing scores of women, despite a personality for which "abrasive" would be a charitable term, for the production of her colossal Dinner Party and somewhat less ambitious Birth Project.[23]

At this writing (fall 2006), the United States is in the grip of a twenty-first-century craze for knitting, and for cozy and sociable independently owned knitting stores, a trend that was not even a cloud on the horizon in the retail census of 1997. This movement, too, has a strong feminist element, although many of its participants would not use this term to describe themselves. The contemporary renaissance of knitting has given rise to knitting circles, now renamed "Stitch 'n' Bitch" clubs to denote their assertive and insouciant departure from needlework traditions. Another strand is the connection with spirituality, adherents of which extol the contemplative virtues of knitting and its proven power to soothe and relax the stressed-out modern psyche.[24]

The Leisure Market for Pleasure: Consumer Motivations

What has been driving this centuries-long upward trend in the market for needle-work and other hedonizing technologies? In Chapter 1, we discussed and, for all practical purposes, dismissed the hypothesis that hobbies are an artifact of modern alienation from work, as this does not explain why Elizabethan noblewomen made preserves, or why young men in the late nineteenth century loved pretending to be eighteenth-century coachmen, to name only two of the many questions not answered by the paradigm set forth by Steven Gelber and others. Some significant human need for worklike play, noted by anthropologist Clifford Geertz, among others, has expanded into the population as leisure has democratized. Clearly, part of the attraction is the closing-out of ordinary daily concerns: the capacity of the hobby to fully engage the attention of the participant seems to be a critical element, as is the calming and self-affirming quality of power over both the process and the product. Knitters, anglers, model railroaders, audiophiles, and other hobby artisans and leisure-activity enthusiasts all report that the immersion of the senses and the intellect into the activity are the main motivation for engaging in it.[25]

This, of course, is not really an explanation; indeed, much of what I shall suggest in this section is descriptive rather than explanatory. For example, the kind of hobbyists described here often are great lovers of tools, especially old ones; it is not unusual for artisans, both amateurs and professionals, to collect antique examples of the tools, designs, and materials of their craft.[26] This explains nothing: why should we love old tools? That many of us do is, however, an undeniable fact, and one with considerable significance to the paths of arts like knitting, in which wooden or bamboo needles have taken over part of the market for their counterparts in aluminum, steel and plastic. Audiophiles are said to mistrust modern digital recording and to prefer older analog technologies like vinyl LPs.

Nostalgia plays a role in some kinds of hobby artisanship. In the 1960s and 1970s, for example, the romanticization of rural life and crafts played a role in their popularity among the young. But this cannot explain the rise of hobbies like ham radio, high-fidelity, photography, antique auto restoration, paintball, and modelmaking. While some models built by amateurs are of the mechanisms of the past, such as sailing ships, others are of futurist or state-of-the-art devices,

Many craft hobbies are consciously archaizing, but scientific hobbies like amateur rocketry are often forward-looking in technological perspective. In this 1973 illustration for Dane Boles' *Parks & Recreation* article "Getting It Off the Ground," Boles presides over an exhibition of junior rocketry. Photo in the collection of Dane Boles, reprinted by permission.

such as automobiles and spacecraft. In the 1950s and 1960s, junior rocketry was a popular hobby among boys; clearly this activity was forward- and not backward-looking in its technological perspective.[27]

As photo-essayists James Twitchell and Ken Ross have pointed out, hobbies may permit their practitioners to withdraw not only from their ordinary concerns, but also from people, such as spouses and other family members, with whom they ordinarily associate.[28] Some hobbyists are happiest when fully absorbed in solitary pursuit of their muse; others prefer to work with like-minded amateurs, either independently or as part of organizations like model train clubs, quilting groups, and "stitch 'n' bitch" circles.

Finally, hobby crafts provide a venue for skill and the pursuit of excellence. This is perhaps the only aspect of hobbies that is consistent with the alienation theory of hobbies: persons who may be entirely apathetic about their work may find the passion of their lives in leisure activities. Yet this is true also of sports, and not just for participants. Sports fans rejoice in their knowledge of the history of their sport and form much the same kind of market for information that hobby artisans do. But hobbies allow a sharing of the product, and uses of it as expressions of pride and love, in a way that sports do not; a cake decorator, woodcrafter, or needleworker can make a gift of his or her artisanship in a way that an athlete or sports fan cannot. According to the 1992 *Nationwide Consumer Study* by the Hobby Industries Association of America, 81 percent of hobbyists crafted objects to use as gifts.[29]

Technical Paths and Market Responses

Sociologist of science Donald Mackenzie, writing in 1984 of technological roads not taken, noted that historians interested in "the effect of social relations" on technology may approach their subjects by "documenting the *contingency* of design, identifying instances where 'things could have been different,' where, for example, the same artifact could have been made in different ways, or differently designed artifacts could have been constructed."[30]

In the case of hobby technologies, the possible paths and "different ways" of making artifacts—the historical and possible future technological paths of the craft—are bewilderingly diverse and may take any conceivable (or inconceivable) direction toward or away from "high" technology. Because these artisans have control of their technology, and purchase its tools, materials, and information sources directly, without the mediation of managers concerned with efficiency and productivity, the markets that address these consumers are primarily concerned with customer satisfaction. In short, manufacturers must constantly adjust their offerings to new expressions and definitions of hedonization, often returning to designs abandoned as "obsolete" decades or even centuries ago. The return of wooden knitting needles, the reappearance of tubes in audiophile equipment, and the renaissance of home preserving are all examples of such reverses of technological course.

Sometimes, the tools and materials of some entirely different activity are transformed into hobby technology and are then marketed for this purpose. The short, flat wooden sticks familiar to elementary-school crafts classes were originally manufactured as handles for frozen desserts—Popsicle sticks—but are now packaged and sold as craft supplies. Pipe cleaners were, of course, originally invented and used to clean pipes, but much longer and more thickly tufted models in a variety of colors are now available for artisan use; the irony of these objects is that I have been told by a pipe smoker that the craft-store version is actually superior to traditional pipe cleaners for cleaning pipes.

As these two examples show, and as my tables have shown, hobby activities support a large array of manufacturing and retailing enterprises, many of which are very small and specialized, surviving by staying close to their consumers,

TABLE 5.3.
Hobby craft participation rates in the United States, 1992

Hobby	Participation rate (% of all hobby artisans)
Cross stitch, embroidery, or crocheting	33
Canvas embroidery	25
Craft sewing	23
Knitting	22
Cake decorating / candy making	19
Wreaths / wall décor	18
Art/drawing	17
Quiltmaking	16
Floral arranging	15

Source: At Our Leisure (Long Island City, NY: Alert Publishing, 1992).
 Note: Percentages do not total 100 because some hobbyists participate in more than one craft.

often by interacting with them at shows, or by sending representatives into retail units to find out what consumers want this year, this month, or this week.[31] Hobby artisans are tinkerers by definition, and as historian of technology Yuzo Takahashi has observed of the Japanese radio industry, "Tinkerers . . . are people who can bridge the two worlds of consumers and producers."[32] Hobbyists and their suppliers can talk to each other in ways that are not available to factory workers and those who make the equipment they use. The manufacturers, publishers, and suppliers of hobby tools, materials, equipment, and information resources have no choice but to listen and adapt if they are to survive in the market for pleasure.

And a very good market it is, too: gardening in the United States, for example, is a market of between $70 and $76 billion annually; and Americans spent $2.5 billion on barbecue grills in 2005. [33] More than half of all U.S. households own cameras and take pictures; sales at photography stores totaled about $12.6 billion in 1995.[34] As of 1992, hobby crafts were a component of leisure in 77 percent of American households, with women dominating most of the popular crafts except woodworking and leatherwork. *Research Alert* reported the highest participation rates among the needle arts (see Table 5.3).[35] As we observed in the tables for hobby, photographic, needlework and gardening retail establishments in Chapter 4, all these markets have grown significantly since World War II.

The household in which I live is wholly owned and controlled by a grey mackerel shorthair member of *Felis cattus*, who selected us from candidates presented to her by the Tompkins County, New York, Society for the Prevention of Cruelty to Animals (SPCA) in 1999. When not supervising her human staff, eating, sleeping, sharpening her claws, washing herself, or inspecting her vast territory from the secure loopholed fortification we humans call a "deck," she enjoys a regal leisure in which play is the most active feature. The experts in ailurian biology and psychology tell us that cats who reign over human households in this way are "juvenized"; those, like our own, who are spays or neuters do not even have sex and parenthood to remind them of their status as adults and thus retain the playful character of kittens all their lives. All of the games our cat plays are, in effect, hedonized forms of hunting: chasing catnip-scented balls, savaging short pieces of plastic foam, leaping from behind furniture to attack the feet of her human companions, and sometimes even chasing real rodents and birds outdoors. Occasionally, she even catches one of these unfortunates, but she eats her prey only rarely, as she is simply not hungry enough. More often, she presents it to us as a gift, possibly as a gesture of *noblesse oblige.*

I cannot say whether the tendency to turn a means of making a living into a form of recreation is cross-species; I do not even have the evidence to show whether or not it is broadly cross-cultural, but the parallels with human leisure are obvious.[36] When work is unnecessary, leisure activities that are work-shaped and worklike in their capacity for challenge, creativity, and absorption are adopted and developed and become significant forces in the marketplace. Even when we must spend a third of our lives working; even when, as the economists tell us, work weeks are reversing their historical trend in the United States and getting longer instead of shorter; even when the average American commute to work is nearly half an hour each way—we somehow still make room in our lives for worklike leisure activities and room in our budgets for their tools, materials, and information resources.

As we have seen, the tendency to hedonize technologies is not new; what has characterized the last century and a half is their extension into nearly all strata of society and into a very broad range of both retail and manufacturing industries. While it is not the case, as some have argued, that the Industrial Revolution created hobbies *ex nihilo*, it was certainly responsible for their democratization, both by expanding individual and family leisure and by bringing down the retail prices of everything hobbyists want and need: needles, yarn, fabric, fishing tackle, guns, garden spades, nursery plants, printed designs, hammers, saws, potters' wheels, cake molds, sugar, canning jars, HO railroad track, sewing machines and other power tools, and thousands of other items we are still learning to see as elements of hedonized technology in an industrial world.

Critics of technology such as Jacques Ellul, Theodore Roszak, René Dubos, and Langdon Winner tell us that modern "technique," as Ellul expresses it, is alienating, dehumanizing, corrupting, and/or subversive of democracy, creativity, and morality.[37] Winner and Ellul assert that its development has spun out of our control, carrying us along willy-nilly in its deterministic race toward catastrophe. Albert Borgmann thinks the "device paradigm" robs us of our natural focus on the hearth, home cooking, music-making, and fly-fishing.[38] Borgmann's position is at least compatible with technological hedonization, except that he seems to imply that the activities he extols are, or should be, among those which "a body is obliged to do," a stricture which would, of course, rob us of our pleasure in them. Summarizing this discourse over technological values, philosopher Don Ihde asked in 1993: "Is technology alienating? Dehumanizing? Or is it fulfilling? And choice multiplying? One whole strand of argumentation about technology has arisen in philosophy between those who see in technological development . . . a diminution of many highly valued traditions, usually associated with community and scale, and those who see in the same developments, the growth of pluralism, cosmopolitanism, and a postmodern context."[39]

Hedonized technologies of the type described in the current book are almost entirely absent from this debate. When mentioned at all, hedonized technologies are usually cast in the context of capitalism's antidotes to alienation, in which hobbies, and indeed leisure activities generally, are portrayed either as opiates of the masses—who, drugged by television and Disneyland into submission to sinister capitalist market forces, stagger into Wal-Mart and A. C. Moore to buy the instruments of their own oppression and social control—or as ineffective and ultimately futile rebellion against these same forces.[40]

In part, this refusal to take hedonized technologies seriously seems to be based on a perception that they are frivolous and economically unimportant, despite the billions of dollars and hours that are lavished on them every year in Western industrial democracies. As we have seen, there is also an element of snobbery: most hobbies are thought to be bourgeois, and so they are, and what of it? As music historian Ronald Radano has observed, the burden is on "the Marxist-oriented critics" to make compelling cases for "middle class alienation" and for the supposed greater social value of so-called high culture versus "bourgeois" aesthetics; the latter will not be made to go away by denying their legitimacy.[41] Karl Marx's contempt for the self-indulgent comforts of life above the poverty level apparently takes no account of the reality that the condition of the bourgeoisie is exactly what most of us humans ardently desire: economic security, time for leisure activities, and the freedom to decide for ourselves what forms of self-actualization are appropriate to us. It is surely irrelevant what a nineteenth-century German economist or even a twentieth-century French philosopher might think of hobbies; the market, which, as we have seen, has been centuries in the making, has flooded heedlessly over them like the tide King Canute ordered not to come in. Whether Jacques Ellul or Karl Marx would have approved of them or not, hedonizing technologies are here to stay.

Biases of Collecting and Connoisseurship

The seventeenth century marks the beginning of significant secondary-source coverage of hobby and leisure artifacts and their technologies as collectibles. While there are a few secondary sources on earlier leisure-artisan textiles, as we have seen, the thin stream of needlework historiography before the twentieth century gathered after 1900 into a flood of studies of seventeenth- and eighteenth-century embroidery in Britain and America, followed by a bimodal surge of interest in nineteenth-century American quilts in the remaining years of the twentieth century. Similar processes characterize the literatures of other crafts and hobbies.

This literature is invaluable for its documentation of textile collections and their artisans, and for the details of materials and stitch techniques it offers, but it has two defects as a body of documentation about leisure needlework: first, the standard of scholarship, especially in works before 1970, is generally quite low. Footnotes are rare, and some works do not even have bibliographies or references to the repositories in which the objects and/or documents described in the text are held. Many illustrate and refer to artifacts only to be found in that Bermuda Triangle of lost textiles, the "Author's Collection."[1]

More problematic, however, is the systematic and nearly always unself-conscious bias of connoisseurship that pervades the literature of textiles and their technology as artifacts. It simply does not occur, for example, to the collectors, connoisseurs, and curators Mildred Davis, Ethel Bolton, Betty Ring, or even Susan Burroughs Swan that a researcher might be more interested in the social context of an object than in its quality and desirability as a work of art, and its suitability for connoisseurship.[2] Thus there are enormous lacunae in the historiography: whole shelves of volumes on eighteenth-century samplers and embroidered pictures, but few serious histories of crochet or tatting; bookcases full of quilt history but next to nothing on the nonfunctional (or even anti-functional) embellishment of potholders, dishtowels, or baby clothes.[3] Even within a medium favored by connoisseurs, such as embroidery, museums and the private collectors who are their donors and supporters exhibit and write about artifacts that are, for the most part, out of the mainstream of hobby—or, as museum curators term it, "amateur"—needlework. This bias affects what we think we know about textile crafts as hobbies in several ways.

First, the survival of all artifacts, including books, reflects a whole series of preservation biases, many of which cannot be known. We speculate, but cannot know for certain, for example, that many more needlework pattern books than are now extant must have once existed and have been worn out and discarded. Embroidery historians Marcus Huish and Averil Colby remark that most of the British samplers that survive date from the seventeenth century and later, even though it is clear that the medium was fully developed by 1650. Earlier examples, they suggest, must have been more numerous than surviving textiles indicate.[4]

Huish and Colby point to the existence of a highly visible but so far incomprehensible preservation bias that continues to leave the historical origins of the sampler a matter of conjecture. Similarly frustrating absences, such as the lack of historical evidence that the potholder existed before the late eighteenth century, are perhaps even more maddening because the entire lack of documentation and examples of these artifacts before 1750 may represent either preservation bias or a real historical absence like that of the wheelbarrow from the ancient world. In both cases, satisfactory explanation is both necessary and impossible. How could something as simple and obviously useful as the potholder have failed to be invented before the eighteenth century?[5] But then, how could such compulsive builders as the Romans have failed to invent the wheelbarrow?[6]

Usually but not always implicit in preservation bias is the possession of creation and preservation resources within the historical chain of custody, or "provenance," as museums call it. Thus we have many more ecclesiastical than domestic textiles from the Middle Ages, not only because the Church could afford to commission designs and buy expensive materials for embroidery and lace, but because the Catholic Church everywhere possessed the human resources and built environment to create, care for, and store the costly vestments, gloves, chasubles, frontals, and so on from one generation to the next, and were less likely than private households to lose their entire investment—pun definitely intended—in raids, wars, and fires.

Superimposed on this pattern of preservation bias are the biases of collecting and connoisseurship, in which the rare is privileged over the commonplace, and aesthetic values are deeply—though usually tacitly—influenced by perceptions of present and/or future economic value. Until fairly recently in the history of collecting, any machine-made artifact was regarded as déclassé, as Steven Gelber has documented.[7] Even among handmade objects, eighteenth-century samplers and embroidered pictures, which are relatively rare, are considered much more prestigiously "collectible," as well as more valuable when sold, than christening dresses, which abound in both private and museum collections.

Rare objects have traditionally been preserved, valued and studied by connoisseurs and curators; plentiful ones usually have not. Stamps were among the first counterexamples, but even they are more valuable when rare, even when the cause of rarity is a printing defect. In this context, examples of historical tools and materials are difficult to find even for the pre- and protoindustrial periods, as most of these were either worn out or discarded long before they had the opportunity to become "collectible." Workboxes, from both before and after the Industrial Revolution, are easier to find in museums than the tools and materials that once filled them, in part because most workboxes are essentially small pieces of handmade furniture—in effect, because they are more "collectible" than tools. In the late 1970s, I tried to document the standardization of American and British knitting needle sizes, a development of the late-nineteenth and early-twentieth centuries but was unable to find adequate evidence in museum collections, because so few institutions collected industrially made needlework tools or the catalogs and instruction books that documented the innovations of standard gauges and needle sizes.

Because "quality" distinctions and discrimination—read "snobbery"—are inherent characteristics of connoisseurship and its public-trust counterpart, art curatorship, the collections that survive in both private and public collections thus reflect a secondary bias that further reduces the surviving evidentiary base by selecting from it only those artifacts and documents deemed worthy of preservation for posterity by the discerning and self-evidently superior judgments of connoisseurs.[8] "Connoisseur" means, after all, "the one who knows."

It is these collecting *cognoscenti* who are responsible for most of the early secondary sources on needlework and other crafts history that began to be published in the nineteenth century.[9] In the case of needlework, most of the early-twentieth-century literature has a characteristic lag of at least fifty years, and often more than a hundred. The reason for this, according to these connoisseur-historians, is that the art of needlework, like the theater and no doubt any number of other equally thriving arts, is perpetually dying. Thus, Marcus Huish lamented in 1900 what he regarded as the inadequacy of "the skilled productions from English schools" such as the Royal School of Art Needlework in London to "fitfully to stir up the dying embers of what was once so congenial an employment to wom-ankind, [which] are no indications of any possibility of needlework regaining its hold on either the classes or the masses."[10]

Candace Wheeler, writing in the United States a little over two decades later, expressed a contempt for Berlin canvas embroidery (what modern Americans would call "needlepoint") that was to become a canonical theme in the literature of needlework connoisseurship. Describing the appearance on the embroidery scene of canvas embroidery worked in brightly colored wools according to a charted pattern early in the nineteenth century, Wheeler remarks that "those who in earlier times were devoted to fine embroidery solaced their idleness with this new work—certainly a poor substitute for the beautiful embroidery of the preced-ing generation."[11] Martha Genung Stearns (1886–1972), a kind of rear guard of the Colonial Revival embroidery movement of the 1896–1925 period, sneered in 1963 that "the last seventy-five years have contributed almost nothing, nationally speaking, to American needlework."[12] Embroidery historian Mildred Davis said in 1969 of the Industrial Revolution in textiles that "in the great flood of expand-ing commerce that followed, the artifacts of the eighteenth century were largely forgotten and continued to lie neglected until near the dawn of the twentieth century."[13]

Susan Swan, the most modern and scholarly of this nostalgic and melancholy choir, picked up the doleful refrain in 1977 in the preface to her history of Amer-ican needlework from 1700 to 1850, based on the collections of the Winterthur Museum in Wilmington, Delaware: "My study started with the objects. At the Henry Francis du Pont Winterthur Museum alone, there are more than six hun-dred pieces of American needlework. By arranging these and other pieces, ac-cording to their techniques and dates, and by allowing for variations in individ-ual abilities, I began to see a pattern in the development of the needle crafts in America. The finest work was done before 1785. Between 1785 and 1825, the work was proficient but not as fine as that done earlier. And from 1825 to 1875, a marked deterioration in needlework skills became apparent."[14]

Like Wheeler and Davis before her, Swan has nothing but scorn for what is known as "Victorian" embroidery.[15] These authors would have felt a kinship with the classicist William Whallon of Michigan State University, who asserted in the mid-1960s, "There was Homer, and then the decline began."[16] Collectors and connoisseurs are necessary for preserving evidence and for drawing public atten-tion to the history of artifacts, but they are unreliable and biased as sources of information. Steven Gelber, however, seems to concur in Swan's judgment, re-marking with the fine contempt of a connoisseur:

Embroidery, and its bastardized offshoots, were a microcosm of Victorian crafts. Technology transformed a traditional artisanal activity by making it faster, easier, and more accessible to a mass market with free time. There was no pretense that the hobbyist stitching a pious motto for the parlor was emulating a tradition of female artistry. She was doing something very much of the moment. She was playing a craft game that demanded no artistic ability and only a modicum of sewing skills. Like the paint-by-number pictures and precut furniture kits that would be popular in the next century, Berlinwork and perforated card projects depended on machines but created the illusion of skill so that the hobbyist had a sense of accomplishment, and something to do to fill idle hours.

Since Gelber is so unfamiliar with needlework and its techniques that he confuses whitework with mending, and crochet with shell work, one can only conclude that he has based these astonishingly patronizing generalizations on the opinions of collectors and curators like Swan.[17] There is certainly nothing to suggest that he has attempted Berlinwork himself, using the kind of patterns available in the period he is denigrating, an experiment recommended to anyone inclined to slight the "modicum of sewing skills" the task requires. The literature of connoisseurship is useful to historians, but it is not critical historiography.

Methodological Notes

My approach to what some will no doubt consider to be very primitive bibliometrics was to search WorldCat, the database and online cataloging utility of the Online Computer Library Center (OCLC), formerly the Ohio College Library Center, on the Library of Congress Subject Heading (LCSH) for each of the leisure crafts I had identified as components of the historical trend toward hedonization.[1] Some of these required searches on two headings; works on ham radio, for example, may be classed as "Radio—Amateurs' Manuals" or "Amateur Radio Stations," or both. Formats were limited to exclude visual materials, musical works, sound recordings, and maps. Each subject search began with the "Early Works to 1800" subhead appended to the LCSH, even in cases where I knew there would be no results, as in that of "High-Fidelity Sound Systems." I included 1800 in my searches for the first decade of the nineteenth century, but for the following decades set temporal delimiters of the form "1811–1820."

WorldCat, the online union catalog of the Online Computer Library Center, with more than 50,000 member libraries in 96 countries at this writing, has bibliographic records for about 70 million titles. A "title" is a bibliographic entity with its own identity; for example, although a serial may run to many volumes, it is all one "title." Not all materials in member libraries appear on WorldCat; most large libraries, including the Library of Congress, have not yet completed the irreducibly labor-intensive and expensive process of converting their old card catalogs to MARC (MAchine Readable Cataloging) format records compatible with modern international bibliographic standards. WorldCat is, however, one of the largest bibliographic databases available to scholars and is readily searchable on a wide variety of access points: subject, author, title, ISBN, and so on.

One of WorldCat's many virtues is that it represents the product of more than a hundred years of argument, counter-argument, and consensus-building among librarians about controlled vocabularies, which makes it possible for researchers to pull up more consistently related titles using LCSHs than would be feasible using keywords. A keyword search on "Needlework," for example, even with date limitations, would yield a bewildering and overwhelming number of references. Such a search would also fail to retrieve works like Pagano's 1550 *Giardineto Novo di Punti Tagliati et Gropposi* because the string "needlework" appears nowhere in the bibliographic record.[2]

Like all databases, WorldCat has limitations and biases. It is dominated by works in English, is weak in ephemera (especially of the "pulp" type), and is limited to what is in libraries and cataloged. Most of the data comes from large repositories like university libraries. Cataloging conventions—most notably, dating works of unknown date by probable century— create statistical artifacts.[3] In the case of a publication of unknown date, the cataloger may only be able to identify or make an educated guess at the century in which the work was published, e.g., "1800s" or "1900s." In the tables and charts at the end of this appendix, this creates a "heaping" effect on the initial years of centuries, creating the false impression that more works were published in these years than in the immediately preceding and following years.[4] This effect is especially noticeable in the case of serials.

Librarians define serials as "works issued or intended to be issued in successive parts bearing numerical or chronological designations and intended to be continued indefinitely"—journals, magazines, newspapers, and so on.[5] Overall, more monographic titles are published than serial titles, and the data set for the monographs is correspondingly larger. Not only does the smaller data set for serials show these heaped titles as a larger component of the total, but serials are typically cataloged from significantly less than a complete run, sometimes from only one or two issues. If the cataloger has in hand, for example, an issue from October 1837 enumerated "Vol. 2 no. 3" and another dated February 1845, he or she can safely assume only that the title began publishing sometime in the nineteenth century, and can assume nothing about when, or even if, it ceased publication. The date range in the catalog record would be recorded as "1800s—"), with the blank representing the unknown date at which the title ceased. Titles cataloged in this way are heaped on the initial years of centuries when the data are sorted by date.

Because cataloging typically includes only initial dates of serials, and dates on which serials ceased publication are not always known, it is difficult or impossible to determine how many serials on a topic were being published in any given year. For both serials and monographs, changes in cataloging conventions over time are reflected in the database: for example, many books on brewing from

1687 on are clearly intended both for professionals and for home brewers, but the LCSH "Brewing—Amateurs' Manuals" does not appear before 1960.

This kind of lag is not unusual; many LC headings appear much later than the literatures they are intended to describe, and a few are underutilized by catalogers thereafter. For example, "Pottery—Amateurs' Manuals" lists far fewer titles of this type than one can find by searching on "Pottery." One LCSH, "Models and Model Making," proved very difficult to work with because after the mid-nineteenth century works of mathematical and statistical model-making appeared, and these are of course included in the same subject heading with physical models such as those of steam engines.

While I always attempt to make my results as reproducible by others as I can, researchers attempting to check my counts of titles will have difficulty doing so because WorldCat, like other major databases, changes daily as works are added to it by thousands of libraries all over the world; titles are sometimes removed, as well, if the only repository (called a "loc" [location], pronounced to rhyme with "choke," in library parlance), withdraws or loses its only copy. Moreover, plans were afoot, in the summer of 2006, to merge OCLC with the Research Libraries Information Network (RLIN), the smaller and older database of the Research Library Group.[6]

On the question of the earlier hedonization of embroidery and lace than any of the other needle arts, a subject search of LCSHs of WorldCat on "Embroidery—Early works to 1800" yields fifteen titles in four languages—English, Italian, German, and Latin—of which the earliest is Pagano's, dated 1550. A similar LCSH search of "Lace—Early Works to 1800" produces only a single title, a translation of Cesare Vecellio's 1617 *Corona delle Nobili et Virtuose Donne*, although if we search "Needlework—Early works to 1800," we find twelve titles in English, German, and Italian, of which some relate to cutwork and other needle-made laces.[7] "Early works" searches of LCSH headings "Crochet," "Tatting," "Knitting," and "Sewing" find no records at all, and there are only three records for early works on "Dressmaking," all of them modern reprints of Juan de Alcega's 1589 *Tailor's Pattern Book*.[8]

See pages 138–46 for the table and figures for this appendix.

Hobby crafts publications, 1500–2000

A. 1500–1900

Library of Congress Subject Heading	Up to 1800	1800–1810[a]	1811–1820	1821–1830	1831–1840	1841–1850	1851–1860	1861–1870	1871–1880	1881–1890	1891–1900[a]
Needlework Crafts											
Needlework: total	1	2	4	4	11	64	46	29	57	124	192
Needlework: serials	0	0	0	0	0	3	12	10	4	6	65
Embroidery: total	16	20	3	0	4	17	18	11	44	84	216
Embroidery: serials	0	1	0	0	1	1	0	0	0	2	15
Knitting: total	0	13	0	0	8	36	9	9	14	46	98
Knitting: serials	0	0	0	0	0	1	0	0	0	3	32
Crocheting: total	0	8	0	0	2	33	7	1	4	38	97
Crocheting: serials	0	0	0	0	0	0	0	0	0	2	32
Lace & lace making: total	34	14	3	2	1	9	8	14	49	59	147
Lace & lace making: serials	0	1	0	0	0	0	0	0	0	1	13
Tatting: total	0	0	0	0	0	2	0	0	0	0	9
Tatting: serials	0	0	0	0	0	0	0	0	0	0	0
Sewing: total	117	117	3	6	4	13	19	19	24	36	127
Sewing: serials	0	5	0	0	0	0	1	0	1	0	35
Clothing & dress—repairing	0	0	0	0	0	0	0	0	1	0	2
Quilts: total	2	1	0	0	0	0	0	3	0	0	51
Quilts: serials	0	0	0	0	0	0	0	0	0	0	40
Other Traditional Crafts											
Canning & preserving: total	69	9	9	5	6	6	8	16	7	35	98
Canning & preserving: serials	0	2	0	0	0	0	0	0	1	0	37
Cookery or Cooking: total	1,763	431	197	292	301	446	513	501	783	1,562	4,917
Cookery or Cooking: serials	1	15	0	0	0	0	2	2	7	7	389
Brewing: total	139	68	37	51	40	52	39	55	67	298	175
Brewing: serials	0	16	0	0	0	1	1	9	105	159	14
Fishing: total	490	105	59	98	131	148	206	234	260	439	1,174
Fishing: serials	0	18	0	0	7	0	0	3	19	7	591
Hunting: total	534	118	83	114	136	197	447	407	560	692	1,931

Hunting: serials	0	17	4	11	13	2	3	3	21	7	932
Shooting: total	17	23	22	23	15	18	74	66	69	91	284
Shooting: serials	0	6	0	0	0	0	0	3	1	2	147
Gardening: total	1,255	234	116	210	263	314	381	295	375	445	1,245
Gardening: serials	3	59	5	11	36	49	35	37	50	54	520
Handicraft: total	13	23	2	8	8	9	19	33	40	86	413
Handicraft: serials	0	5	0	0	0	1	1	0	2	7	272
Woodworking: total	1	11	0	1	0	3	3	12	31	71	147
Woodworking: serials	0	2	0	0	0	0	0	0	0	0	53
Carpentry: total	95	64	39	47	73	71	87	62	82	83	122
Carpentry: serials	1	0	0	1	0	1	1	0	4	1	815
Turning: total	5	3	6	3	2	10	7	17	11	21	24
Turning: serials	0	0	0	0	0	0	0	0	0	0	0
Pottery: total	13	32	4	5	17	31	43	87	224	198	422
Pottery: serials	0	5	0	0	0	0	0	2	13	3	73
Modern Crafts											
Radio, amateurs' manuals: total	0	0	0	0	0	0	0	0	0	0	0
Radio, amateurs' manuals: serials	0	0	0	0	0	0	0	0	0	0	0
Amateur radio stations: total	0	0	0	0	0	0	0	0	0	0	0
Amateur radio stations: serials	0	0	0	0	0	0	0	0	0	0	0
Lithography: total	0	24	49	99	106	61	70	49	35	33	131
Lithography: serials	2	2	0	1	0	3	2	2	3	3	15[b]
Leatherwork: total	1	1	0	0	0	3	5	0	2	4	20
Leatherwork: serials	0	0	0	0	0	0	0	0	0	0	4[b]
Sailing: total	4	4	1	3	9	13	35	39	59	86	60
Sailing: serials	0	1	0	0	0	0	1	0	1	0	29[b]
Boats and boating: total	12	16	1	6	14	24	33	44	38	63	405
Boats and boating: serials	0	6	0	0	0	0	0	0	0	0	286[b]
Camping: total	16	3	0	0	1	1	13	25	48	94	182
Camping: serials	6	1	0	0	1	1	0	0	0	0	96
Photography: total	3	93	2	2	77	186	509	443	355	755	1,909
Photography: serials	1	58	0	0	3	10	73	45	32	85	715[b]

[a]As noted in the text of the appendix, "heaping" for the initial years of centuries and decades—i.e. an artificially high number of publications—is an artifact of cataloging conventions for works of uncertain dates of publication.

[b]Heaped data.

(continued)

Hobby crafts publications, 1500–2000 (cont.)

B. 1901–2000

Library of Congress Subject Heading	1901–1910	1911–1920	1921–1930	1931–1940	1941–1950	1951–1960	1961–1970	1971–1980	1981–1990	1991–2000
Needlework Crafts										
Needlework: total	81	98	82	112	71	107	215	656	699	596
Needlework: serials	5	2	6	5	5	5	11	20	23	23
Embroidery: total	147	159	170	153	120	231	409	841	1103	1,322
Embroidery: serials	0	5[b]	3	1	1	2	3	5	1	9
Knitting: total	18	65	28	66	122	93	232	610	1174	630
Knitting: serials	1	1	3	1	2	5	3	11	19	10
Crocheting: total	23	125	22	87	102	63	109	574	622	731
Crocheting: serials	0	2	1	1	2	3	4	12	14	11
Lace & lace making: total	126	98	99	65	67	67	89	165	356	4
Lace & lace making: serials	3	3	1	0	1	0	0	1	4	5
Tatting: total	4	17	0	7	24	12	14	41	47	82
Tatting: serials	0	1	0	0	0	0	0	0	1	0
Sewing: total	52	154	149	133	170	201	369	839	748	724
Sewing: serials	0	0	2	0	2	2	6	9	14	11
Clothing & dress— repairing	0	0	5	3	19[c]	2	12	20	14	9
Quilts: total	4	5	9	10	11	8	22	158	695	1,678
Quilts: serials	0	0	0	0	0	0	1	13	20	38
Other Traditional Crafts										
Canning & preserving: total	69	321	193	239	358	232	192	403	356	308
Canning & preserving: serials	6	17	8	9	12	15	10	11	5	0
Cookery or Cooking: total	2,895	3,474	3,849	4,770	5,170	6,967	12,000	26,948	36,624	48,230
Cookery or Cooking: serials	11	10	17	34	45	41	42	159	240	272
Brewing: total	133	210	160	205	278	501	774	923		
Brewing: serials	91	40	19	26	24	32	60	94		
Fishing: total	478	318	504	669	794	1,137	1,713	3,131	3,987	4,740
Fishing: serials	11	11	24	65	75	89	124	190	204	179
Hunting: total	990	787	1,250	1,151	1,021	1,665	2,254	3,544	6,170	9,206
Hunting: serials	12	15	28	59	65	101	136	267	389	508

	1	2	3	4	5	6	7	8	9	10
Shooting: total	135	203	128	160	172	250	495	692	738	900
Shooting: serials	7	3	5	3	8	19	30	30	37	41
Gardening: total	860	1,201	1,092	1,319	1,407	1,526	1,854	5,032	5,833	8,850
Gardening: serials	69	78	109	107	110	115	108	241	275	243
Handicraft: total	129	180	255	499	598	812	1,421	4,097	4,061	6,818
Handicraft: serials	7	2	14	19	22	56	45	163	177	148
Woodworking: total	58	90	104	90	95	171	220	479	487	117
Woodworking: serials	0	1	3	5	9	19	13	18	27	516
Carpentry: total	104	87	90	88	164	152	195	472	469	21
Carpentry: serials	2	1	5	1	3	1	1	3	3	3
Turning: total	28	29	11	16	38	44	50	74	130	237
Turning: serials	0	0	0	0	0	0	0	0	6	3
Pottery: total	462	361	577	642	588	861	1,421	2,392	2,515	2,702
Pottery: serials	13	11	32	11	29	18	19	42	38	38
Modern Crafts										
Radio, amateurs' manuals: total	0	0	15	5	20	80	135	181	178	183
Radio, amateurs' manuals: serials	0	0	3	0	2	5	3	5	15	8
Amateur radio stations: total	0	1	7	12	8	26	81	158	231	256
Amateur radio stations: serials	0	0	1	2	1	3	8	12	27	13
Lithography: total	63	114	147	150	147	163	345	328	345	336
Lithography: serials	3	3	3	3	8	2	9	3	2	2
Leatherwork: total	16	23	48	55	73	67	65	195	132	111
Leatherwork: serials	0	0	0	0	1	1	1	1	2	0
Sailing: total	31	12	45	123	108	155	282	634	616	867
Sailing: serials	3	0	2	1	2	2	12	19	28	14
Boats and boating: total	130	152	128	212	183	369	760	1,997	2,001	2,700
Boats and boating: serials	11	1	3	7	23	30	54	107	137	126
Camping: total	141	242	320	285	330	488	696	1,119	975	1,373
Camping: serials	3	4	14	10	11	20	30	61	52	69
Photography: total	872	607	708	1,198	1,164	1,678	2,422	5,099	5,626	7,389
Photography: serials	88	51	81	139	122	157	166	285	291	241

[a] Heaping for the initial years of centuries and decades is an artifact of cataloging conversions for works of uncertain dates of publication.
[b] Heaped data.
[c] Of these 19 titles, 15 were published between 1941 and 1946.

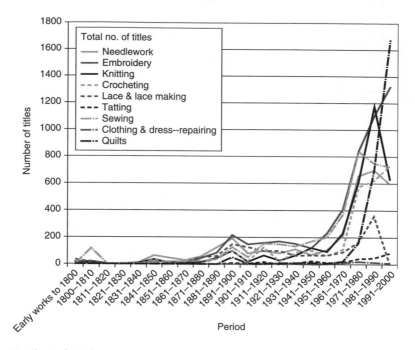

Textile Crafts Titles, 1500–2000

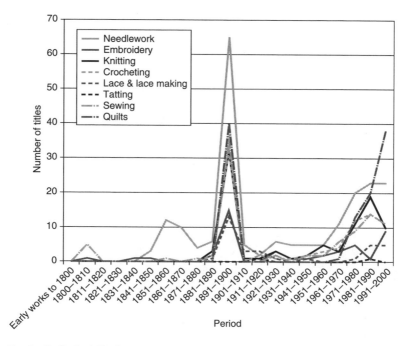

Textile Crafts Serial Titles, 1500–2000

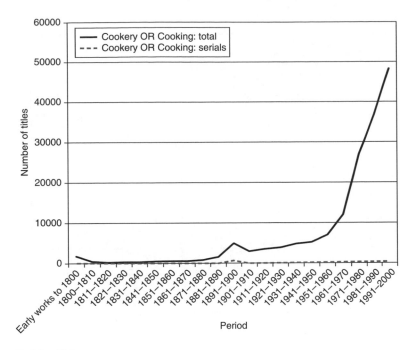

Cooking Titles, 1500–2000

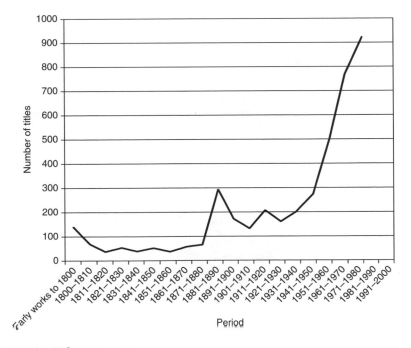

Brewing Titles, 1500–2000

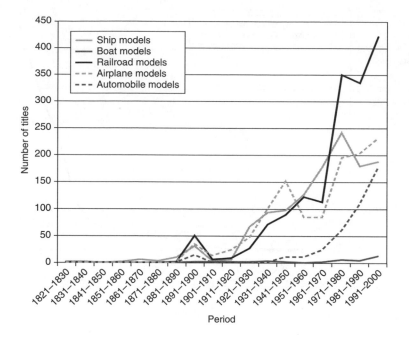

Model-making Titles, 1821–2000

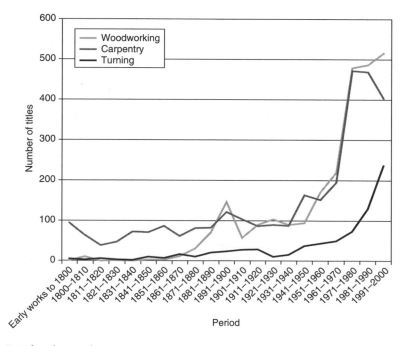

Woodworking Titles, 1500–2000

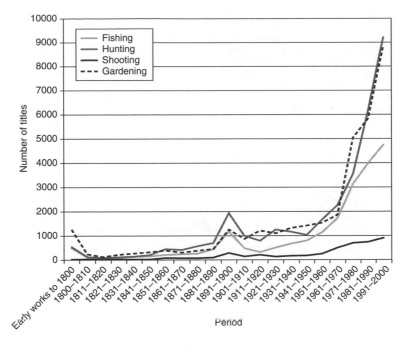

Fishing, Hunting, and Gardening Titles, 1500–2000

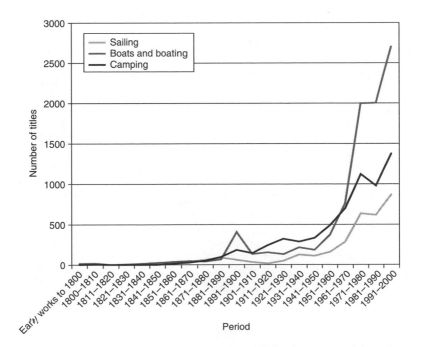

Sailing, Boating, and Camping Titles, 1500–2000

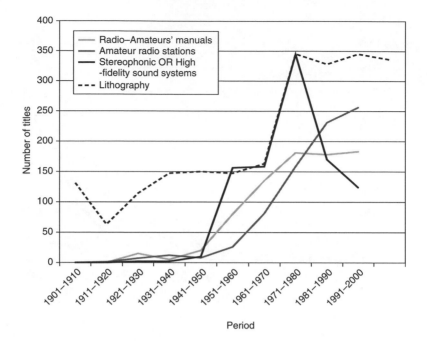

Hedonized Communication Titles, 1901–2000

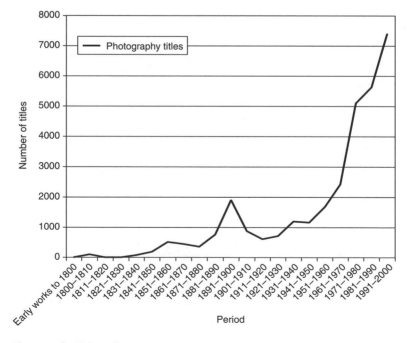

Photography Titles, 1821–2000

Introduction

1. J. C. Horn, "It Says Here Wives Prefer Reading to Sex," *Psychology Today* 12 (March 1979): 17; Jay Mancini and Dennis Orthner, "Recreational Sexuality Preferences among Middle-Class Husbands and Wives," *Journal of Sex Research* 14, no. 2 (1978): 96–106. For more recent results, see *At Our Leisure* (Long Island City, NY: Alert Publications, 1992).

2. Christina Holmes, editorial, *Sew Business*, April 1979.

3. "Leisure Time Activities," *Gallup Opinion Index*, report no. 146 (September 1977). 14–15. Oddly, a July 1991 study of leisure activities by the same organization does not so much as mention sewing or needlework. *A Gallup Study on Leisure Activities* (Princeton, NJ: Gallup, 1991).

4. This possibility is explored in Thorstein Veblen, *The Instinct of Workmanship and the State of the Industrial Arts* (New Brunswick: Transaction Publishers, 1990).

5. Rosalynn Voaden, "All Girls Together: Community, Gender, and Vision at Helfta," in *Medieval Women in Their Communities*, ed. D. Watt (Toronto: University of Toronto Press, 1997), 76; Shulamith Shahar, *The Fourth Estate* (London: Methuen, 1983), 44.

6. There is a modern fictional reference to nuns' enjoyment of embroidery in Maura Laverty, *Never No More* (London: Longmans, Green, 1942), 164.

Chapter 1 · What Is a Hedonizing Technology?

1. For the element of spiritual pleasure in such pastimes, see Kathleen Norris, *The Quotidian Mysteries: Laundry, Liturgy, and "Women's Work"* (New York: Paulist Press, 1998).

2. Diane Schoemperlen, *Our Lady of the Lost and Found* (New York: Penguin, 2002), 319, 322. See also Elizabeth Berg *The Art of Mending* (New York: Ballantine, 2004), 14; and Amy Wilentz, *Martyrs' Crossing* (New York: Ballantine, 2001), 106, in which the chief female protagonist irons her dead son's clothes as part of the grief process.

3. Samuel C. Florman, *The Existential Pleasures of Engineering* (New York: St. Martin's Press, 1976), 127–52. There is another fictional reference to this kind of hypnotic pleasure in crafts in Sue Grafton, *S is for Silence* (New York: G. P. Putnam's Sons, 2005), 54.

4. There is a humorous fictional account of this feature of camping in Elizabeth Von Arnim, *The Caravaners* (1930; London: Virago, 1989), 144–58.

5. Mintel International Group, Ltd., *Outdoor Barbecue—U.S.* (London: Mintel, 2005); on Australians see Mark Thomson, *Meat, Metal, & Fire* (New York: HarperCollins, 1999).

6. John F. Kasson, *Amusing the Millions* (New York: Hill & Wang, 1978).

7. Ruth Schwartz Cowan, *More Work for Mother* (New York: Basic Books, 1983), 149–50. See also a fictional reference to this phenomenon in Anthony Trollope, *Castle Richmond* (1860; London: Penguin, 1993), 144.

8. *Future Food Trends,* marketing newsletter of Technomic, Inc. (December 2004); see also Jerry Adler et al., "Takeout Nation," *Newsweek* 143, no. 6 (2004): 52.

9. Rachel Maines, *Food Service and Retail* (New York: Packaged Facts/MarketResearch .com, 2005).

10. Reese Jenkins, *Images and Enterprise* (Baltimore: Johns Hopkins University Press, 1975), 115–20, 236–45.

11. There are exceptions; see, for example, Kenneth P. Czech, *Snapshot* (Minneapolis: Lerner Publications, 1996); Robert C. May, *The Lexington Camera Club, 1936–1972* (Lexington: University of Kentucky Art Museum, 1989); and Grace Seiberling and Carolyn Bloore, *Amateurs, Photography, and the Mid-Victorian Imagination* (Chicago: University of Chicago Press, 1986).

12. Examples of this type of "mainstream" photographic history that focus on professionals include William Crawford, *The Keepers of Light* (Dobbs Ferry, NY: Morgan & Morgan, 1979); Jack Howard Roy Coote, *The Illustrated History of Colour Photography* (Surbiton, UK, 1993); and Elspeth Brown, *The Corporate Eye* (Baltimore: Johns Hopkins University Press, 2005).

13. Robert C. Post, *High Performance* (Baltimore: Johns Hopkins University Press, 1994).

14. Walter Vincenti, "The Retractable Airplane Landing Gear and the Northrop 'Anomaly': Variation-Selection and the Shaping of Technology," *Technology and Culture* 35, no. 1 (1994): 1–33. I am indebted to the late Marta Bohn-Meyer of NASA for the information on fixed gear in aerobatics.

15. Anne L. Macdonald, *No Idle Hands* (New York: Ballantine Books, 1988).

16. W. Outram Tristam, Herbert Railton, and Hugh Thomson, *Coaching Days and Coaching Ways* (London: Macmillan, 1893); Edmund Vale and Thomas Hasker, *The Mail-Coach Men of the Late Eighteenth Century* (London: Cassell, 1960); Anthony Burgess, *Coaching Days of England* (London: Elek, 1966); and Francis M. Ware, *Driving* (New York: Doubleday, Page, 1903). See also William Makepeace Thackeray, *The Newcomes* (1854; Ann Arbor: University of Michigan Press, 1996), 138. Anthony Trollope, an admirer of Thackeray, refers to coaching nostalgia in his *Dr. Thorne* (1858; Ware: Wordsworth Classics, 1994), 159–60.

17. James B. Twitchell and Kenneth Charles Ross, *Where Men Hide* (New York: Columbia University Press, 2006), 129–46.

18. Archibald Standish Hartrick, *Lithography as a Fine Art* (London: Oxford University Press, 1932); Maurice Bloch, *Tamarind* (Baltimore: Garamond/Pridemark Press, 1971), 5–13.

19. See, for example, Thelma R. Newman, *Creative Candlemaking* (New York: Crown Publishers, 1972); Sandie Lea, *The Encyclopedia of Candlemaking Techniques* (Philadelphia:

Running Press Book Publishers, 1999); and Sue Heaser, *The Big Book of Candles* (Cincinnati: North Light Books, 2002).

20. See, for example, "Shave with Germanic Precision," in *Siegler & Co. Holiday Preview 2005* (West Chester, OH: Siegler & Co, 2005), 7; and the description of "pure badger brushes" on *The Art of Shaving* Web site, www.theartofshaving.com/taos/classic.php (accessed 11 May 2006).

21. I had originally employed the more trendy term "trajectory" rather than "paths" in this context, but was persuaded by anonymous referee that the former term implies force and gravity, neither of which are at work here.

22. Gertrude Whiting, *Tools and Toys of Stitchery* (New York: Columbia University Press, 1928); Sylvia Groves, *The History of Needlework Tools and Accessories* (Feltham, UK: Country Life Books, Hamlyn Pub. Group, Ltd., 1966) ; Molly G. Proctor, *Needlework Tools and Accessories* (London: Batsford, 1990); Bill Stankus, *Setting Up Your Own Woodworking Shop* (New York: Sterling, 1993); Charles R. Self and David Thiel, *The Insider's Guide to Buying Tools* (Cincinnati: Popular Woodworking Books, 2000); and Rodney Frost, *The Nature of Woodworking* (New York: Sterling, 2000).

23. "Star of India Fountain Pen," advertised on *Levenger: Tools for Serious Readers,* www.levenger.com (accessed 11 May 2006).

24. Douglas Martin, "Shelby Foote, Historian and Novelist, Dies at 88," *New York Times,* 29 June 2005, Books section.

25. *Oxford English Dictionary,* 2nd ed. (online ed., 2000), s.vv. "-ize" and "-izer."

26. Abraham H. Maslow, *Motivation and Personality,* 2nd ed. (New York: Harper & Row, 1970), 1–58.

27. A well-observed fictional account of the agonization of knitting is Anne Bartlett, *Knitting* (Boston: Houghton Mifflin, 2005).

28. Clifford Geertz, *The Interpretation of Cultures* (New York: Basic Books, 1973).

29. Thorstein Veblen and Charles Almy, *The Theory of the Leisure Class* (New York: Macmillan, 1899).

30. Leslie Bella, "Women and Leisure: Beyond Androcentrism," in *Understanding Leisure and Recreation,* ed. E. L. Jackson and T. L. Burton (State College, PA: Venture Publishing, 1989), 151–79; Lorine Pruette, *Women and Leisure* (New York: E. P. Dutton, 1924); and Sebastian de Grazia, *Of Time, Work, and Leisure* (New York: Vintage Books, 1994).

31. Euston Quah, *Economics and Home Production* (Aldershot, Hants, UK, and Brookfield, VT: Avebury and Ashgate Publishing Co., 1993), 2, 12, 27, 33, 80, 82, 88, 109, 146, 193, 221. Quah mentions gardening and home sewing but not fancy needlework as hobbies. For strenuous but ultimately unavailing efforts to fit hobby needlework into a home production model, see F. Thomas Juster and Frank P. Stafford, *Time, Goods, and Well-Being* (Ann Arbor: Survey Research Center Institute for Social Research, University of Michigan, 1985); John P. Robinson, *How Americans Use Time* (New York: Praeger, 1977); and the same author's *How Americans Used Time in 1965* (Ann Arbor: Institute for Social Research, University of Michigan, 1977). Note the distinction between "hobbies" and "ladies' hobbies" on p. 84 of this last title.

32. Joffre Dumazedier, *Toward a Society of Leisure* (New York: Free Press, 1967).

33. Fanny Burney, *Evelina* (London: T. Lowndes, 1778); Margaret E. Atwood, *The Blind*

Assassin (Toronto: Random House Canada, 2001), 101; and James Joyce, *Finnegans Wake* (New York: Viking Press, 1947), 28.

34. See, for example, Anthony Trollope, "The Courtship of Susan Bell," in *Complete Shorter Fiction*, ed. J. Thompson (New York: Carroll & Graf, 1992; originally published in *Harper's New Monthly*, August 1860), 22; Margaret Oliphant, *Agnes* (Leipzig: B. Tauchnitz, 1865), 1:1; and Oliphant, *Harry Joscelyn* (Leipzig: B. Tauchnitz, 1881) 1:16–17, 169–70. See also Charlotte Mary Yonge, *The Pillars of the House* (Leipzig: B. Tauchnitz, 1873) 1:280, 2:268.

35. "How to Buy Yarn for Crocheting," *E-How*, www.ehow.com/how_7417_buy-yarn-crocheting.html (accessed 11 May 2006); "The Most Beautiful Color in the World" *Grumperina Goes to Local Yarn Shops and Home Depot*, www.grumperina.com/knitblog/archives/2006/04/the_most_beauti.htm (accessed 11 May 2006). On hand-painted yarns and their uses, see Maie Landra, *Knits from a Painter's Palette* (New York: Sixth & Spring Books, 2006). See also the fictional account of buying yarn in Annie Proulx, *Postcards* (New York: Scribner, 1995), 177–78.

36. Steven M. Gelber *Hobbies: Leisure and the Culture of Work in America* (New York: Columbia University Press, 1999), 1–6. In his discussion of nineteenth-century needlework (pp. 157–80), Gelber does not distinguish between crewel embroidery and crochet (p. 167).

37. Rhona and Robert N. Rapoport, "Four Themes in the Sociology of Leisure," *British Journal of Sociology* 25, no. 2 (1974), 215–29.

38. Lucius Annaeus Seneca, *De Otio*; Margaret H. Swain, *The Needlework of Mary, Queen of Scots* (New York: Van Nostrand Reinhold, 1973); and Izaak Walton, *The Compleat Angler; or, The Contemplative Man's Recreation* (London: Printed by T. Maxey for Rich. Marriot, 1653). On the challenges of applying Marxist principles to the ancient world, see Paul Anthony Cartledge and David Konstan, "Marxism and Classical Antiquity," *Oxford Classical Dictionary*, 3rd ed. (Oxford: Oxford University Press, 1996), 933–34.

39. Hugh Cunningham, *Leisure in the Industrial Revolution, c. 1780–c. 1880* (New York: St. Martin's Press, 1980), 119. The subject of British working-class women's hobby needlework and its documentation is addressed in Chapter 4.

40. Italics in the original. Jennifer Hargreaves, "The Promise and Problems of Women's Leisure and Sport," in *Leisure for Leisure*, ed. C. Rojek (New York: Routledge, 1989), 147. In regard to the book title here, American readers are reminded that Britons pronounce "leisure" to rhyme with "measure."

41. Harold Wilensky, "Work, Careers, and Social Integration," *International Social Sciences Journal* 12 (1960): 543–60; Stanley Robert Parker, *The Sociology of Leisure* (New York: International Publications Service, 1976); and Parker, *The Future of Work and Leisure* (London: MacGibbon and Kee, 1971).

42. John Clarke and Charles Critcher, *The Devil Makes Work* (London: Macmillan, 1985), 3.

43. Homer, *Odyssey*, bk. 7, lines 112ff; Linda Farrar, *Ancient Roman Gardens* (Stroud: Sutton, 1998); and R. J. Forbes, *Studies in Ancient Technology*, vol. 4 (Leiden: E. J. Brill, 1955).

44. Raoul d'Harcourt, Grace G. Denny, and Carolyn M. Osborne, *Textiles of Ancient Peru and Their Techniques* (Seattle: University of Washington Press, 1974).

45. Hans Jürgen Hansen, *Art and the Seafarer* (London: Faber, 1968), 281; *The Naval and Marine Collection of the Late Lieut. Commander William Barrett R.N., London, England* (New York: Anderson Galleries, 1926); and P. W. Blandford, "Decorative Rope and Canvas Work," in *The Decorative Arts of the Mariner*, ed. G. Frere-Cook (Boston: Little Brown, 1966).

46. Bell Irvin Wiley, *The Life of Johnny Reb* (Indianapolis: Bobbs-Merrill, 1943), 170; and Michael P. Gray, *The Business of Captivity* (Kent, OH: Kent State University Press, 2001).

47. Some makers of heating equipment in the early twentieth century advertised this potential for making use of home spaces previously too cold or dirty in the winter, especially when gas began to replace coal as a heating fuel; an example was the 1920 brochure by the International Heater Company of Utica, NY, *International Onepipe Heater Makes Your Entire Home Warm, Cozy and Comfortable; Keeps the Cellar Cool.*

48. Johan Huizinga, *Homo Ludens* (Boston: Beacon Press, 1955), 166.

49. De Grazia, *Of Time, Work, and Leisure,* 69–79, 97–108, 450–51. Although he clearly sees sewing as analogous to home carpentry, de Grazia cites, without comment, a study of leisure activities, "Economic Status and Interests of College and Non-College Women, October 1956," which does not so much as mention needlework, cooking, or gardening.

50. Ruth Schwartz Cowan, "The Consumption Junction: a Proposal for Research Strategies in the Sociology of Technology," in *The Social Construction of Technological Systems,* ed. W. E. Bijker, T. P. Hughes, and T. J. Pinch (Cambridge: MIT Press, 1987), 261–80.

51. Mihaly Csikszentmihalyi, *Beyond Boredom and Anxiety* (San Francisco: Jossey-Bass, 1975), 11–23, 140–60.

52. J. P. Toner, *Leisure and Ancient Rome* (Cambridge, UK: Polity Press, 1995), 18.

53. Roosevelt Grier, *Needlepoint for Men* (New York: Walker and Co., 1973). The threat of retaliation against aspersions on his masculinity might have been implied by Grier's size, but in fact he has no record of violence off the football field. See Roosevelt Grier and Dennis Baker, *Rosey, an Autobiography: The Gentle Giant* (Tulsa: Honor Books, 1986).

54. Roger Hensley, "Interview with Master Model Railroader Mary Miller, MMR," *NMRA: National Model Railroad Association,* 1 January 2001, www.nmra.org/beginner/mmr_mary.html. For fictional male needleworkers, see Wilkie Collins, *The Legacy of Cain* (1888; Dover, NH: Alan Sutton, 1993), 314; the character of Miserrimus Dexter in the same author's 1875 *The Law and the Lady* (Oxford: Oxford University Press, 1992), 235–36; and E. H. Young, *Chatterton Square* (1947; London: Virago, 1987), 200. In Antonia White's 1952 *The Sugar House* (repr., London: Virago, 1979), the gay male actor Trevor is "an expert at embroidering" (p. 29). In this example, the character's homosexuality seems to be performing the rhetorical work of an infirmity.

55. On the relationships among knotting, netting, and tatting, see Groves, *History of Needlework Tools and Accessories,* pp. 78-89.

56. Elgiva Nichols, *Tatting* (New York: Dover, 1984).

57. James H. Gilmore and Joseph Pine, *Markets of One* (Boston: Harvard Business School Press, 2000).

58. Some of these technologies are not all that much less dangerous than their adult counterparts. See, for examples of these, Conn and Hal Iggulden, *The Dangerous Book for Boys* (New York: HarperCollins, 2007).

59. Svante Lindqvist, "Meccano," presentation at annual meeting of the Society for the

History of Technology, Cleveland, OH, Western Reserve Historical Society, 1990. On looper looms, see Barbara Kane, *Potholders and Other Loopy Projects* (Palo Alto, CA: Klutz, 2003).

60. Yuzo Takahashi, "A Network of Tinkerers: The Advent of the Radio and Television Receiver Industry in Japan," *Technology and Culture* 41 no. 3 (2000): 460–84.

61. On hedonizing equestrianism and vehicle driving, see, for example, Vladimir S. Littauer, *Be a Better Horseman* (New York: Derrydale Press, 1941), and Ware, *Driving*.

Chapter 2 · Leisure and Necessity

1. C. M. Woolgar, *The Great Household in Late Medieval England* (New Haven: Yale University Press, 1999), especially p. 121 on fishing. See also Richard C. Hoffman, "Fishing for Sport in Medieval Europe: New Evidence," *Speculum* 60, no. 4 (1985), 877–902; and William H. Forsyth, "The Medieval Stag Hunt," *Metropolitan Museum of Art Bulletin*, n.s., 10, no. 7 (1952), 203–10.

2. The significance of this phenomenon did not, however, escape British feminist historian Rozsika Parker. See her 1984 book *The Subversive Stitch* (London: Women's Press), 25–59.

3. See, for example, Judith M. Barringer, *The Hunt in Ancient Greece* (Baltimore: Johns Hopkins University Press, 2001), 42–46.

4. Pliny, *Historia naturalium*, bk. 19, chap. 19, sec. 4.

5. The Hungarian novelist Sándor Márai (1900–1989) mentions "a huntsman, but only in the way of someone bowing to social customs" in *Embers* (original publication, Budapest, 1942), trans. Carol Brown Janeway (New York: Vintage, 2002), 122. On medieval hunting see John M. Carter, "Sport, War, and the Three Orders of Feudal Society: 700–1300," *Military Affairs* 49, no. 3 (1985), 132–39; Thomas T. Allsen, *The Royal Hunt in Eurasian History* (Philadelphia: University of Pennsylvania Press, 2006), 2–6, and the section on "pursuing pleasures," pp. 193–97; and Dodgson H. Madden, *A Chapter of Mediaeval History: The Fathers of the Literature of Field Sport and Horses* (London: J. Murray, 1924).

6. E. J. W. Barber, *Women's Work* (New York: Norton, 1994), chap. 1. On the care of the sick, see, for example, the fictional account in Theodore Dreiser, *Jennie Gerhardt* (1911; Oxford: Oxford University Press, 1991), 296. For a fictional portrayal of child care and knitting, Charlotte M. Yonge, *A Reputed Changeling* (Leipzig: B. Tauchnitz, 1890), 1:265–66.

7. See, for example, the fictional account of maritime mending in Arturo Pérez-Reverte, *The Nautical Chart*, trans. M. S. Peden (New York: Harcourt, 2001), 270.

8. Susan B. Swan, *Plain and Fancy* (New York: Holt Rinehart and Winston, 1977), chap. 1; Penelope Byrde, *Jane Austen Fashion* (Ludlow: Excellent Press, 1999), 103–18; and Mary E. Jones, *A History of Western Embroidery* (London: Studio Vista and Watson-Guptill Publications, 1969), 7–10.

9. Sarah J. B. Hale, *The Workwoman's Guide* (London: Simpkin, Marshall and Co., 1838), 1–11. For the distinctions between plain and fancy hemming, see Ellen J. Gehret, *This Is the Way I Pass My Time* (Birdsboro, PA: Pennsylvania German Society, 1985), 24–25, 39–44.

10. Martin K. Foys et al. *The Bayeux Tapestry*, digital ed. (Leicester, UK: Scholarly Digital Editions, 2003). On the issue of labor, see Jones, *History of Western Embroidery*, p. 26.

11. Jennifer C. Ward, *English Noblewomen in the Later Middle Ages* (London: Longman, 1992), 70–82; and Shulamith Shahar, *The Fourth Estate* (London: Methuen, 1983), 152–53. On mealtime obligations, see Ffiona Swabey, *Medieval Gentlewoman* (New York: Routledge, 1999), 11, where the author reports that Alice de Bryene's accounts show that in a single year, between 1412 and 1413, "Alice served over 16,500 meals at Acton, an average of 45 meals a day."

12. Sylvia L. Thrupp, *The Merchant Class of Medieval London, 1300-1500* (Ann Arbor: University of Michigan Press, 1962), 170–71; Peter Fleming, *Family and Household in Medieval England* (Basingstoke, Hampshire, UK: Palgrave, 2001), 43, 96; and Barbara A. Hanawalt, "At the Margin of Women's Space in Medieval Europe," in *Matrons and Marginal Women in Medieval Society*, ed. R. Edwards and V. L. Ziegler (Woodbridge, Suffolk, UK: Boydell & Brewer, 1995), 6.

13. Shahar, *Fourth Estate*, 53; Erika Uitz, *The Legend of Good Women* (Mount Kisco, NY: Moyer Bell Ltd., 1990), 172–73; and Penelope Galloway, "'Discreet and Devout Maidens': Women's Involvement in Beguine Communities in Northern France, 1200–1500," in *Medieval Women in Their Communities*, ed. D. Watt (Toronto: University of Toronto Press, 1997), 95. See also Walter Simons, *Cities of Ladies* (Philadelphia: University of Pennsylvania Press, 2001), 85–86, in which Simons seems uncertain whether Beguines embroidered.

14. The consistent gender imbalance throughout the hierarchy of medieval households, documented in account books, remains a mystery to historians. Why does the evidence show so few women? See, for example, Swabey, *Medieval Gentlewoman*, 115–31; Fleming, *Family and Household in Medieval England*, 75; and Woolgar, *Great Household*, 8, 34. On the expensive medieval hobby of falconry for both men and women, see Hans J. Epstein, "The Origin and Earliest History of Falconry," *Isis* 34, no. 6 (1943): 497–509; and Richard Grassby, "The Decline of Falconry in Early Modern England," *Past and Present* 157 (1997): 37–62. Grassby (p. 55) cites the seventeenth-century author Richard Braithwaite to the effect that the costs of falconry were more than eight times the value of the quarry taken by it.

15. A. G. I. Christie, *English Medieval Embroidery* (Oxford: Clarendon Press, 1938), 1–2, 31–33.

16. Shahar, *Fourth Estate*, 44. On distilling in monasteries, Barrington Moore Jr., "Austerity and Unintended Riches," *Comparative Studies in Society and History* 29, no. 4 (1987): 809.

17. Thomas of Ely (fl. 1174), *Liber Eliensis*, ed. Richard Fitzneale, and E. O. Blake (London: Offices of the Royal Historical Society, 1962). See also Christie, *English Medieval Embroidery*, 1–33; and Jones, *History of Western Embroidery*, 11–17.

18. Shahar, *Fourth Estate*, 44. There is a modern fictional account of the pleasures of embroidery in convents in Antonia White, *Frost in May* (New York: Viking Press, 1934), 145.

19. Kay Staniland, *Embroiderers* (Toronto: University of Toronto Press, 1991), 8.

20. *The Ancrene Riwle (The Corpus Ms. Ancrene Wisse)*, trans. M. B. Salu (Notre Dame, IN: University of Notre Dame Press, 1956), 187. On p. 185 we are informed that anchoresses may keep cats, but no other animal. Apparently the subtle joys of dozing off to the dulcet tones of feline snoring were considered less threatening to the soul than those of

unauthorized embroidery and lace. On embroidery as an excuse for truancy from Mass, see Parker, *Subversive Stitch*, 38.

21. Irena Turnau, "The Diffusion of Knitting in Medieval Europe," in *Cloth and Clothing in Medieval Europe*, ed. E. M. Carus-Wilson, N. B. Harte, and K. G. Ponting (London and Edington: Heinemann Educational Books and Pasold Research Fund, 1983), 377.

22. Hoffman, "Fishing for Sport," 899.

23. Sylvia Groves, *The History of Needlework Tools and Accessories*, Country Life Books (Feltham, UK: Hamlyn Publishing Group, 1966), 17.

24. On needles in prehistory and classical antiquity, see S. A. Semenov, *Prehistoric Technology* (New York: Barnes & Noble, 1964), 100–103; R. J. Forbes, *Studies in Ancient Technology* (Leiden: E. J. Brill, 1955), 4:178–79; and Henry Hodges, *Technology in the Ancient World* (New York: Knopf, 1970), 26.

25. Marion Channing asserts that the Romans introduced bone needles into Britain in her 1969 study *The Textile Tools of Colonial Homes* (Marion, MA: Channing), 55.

26. Georgiana B. Harbeson, *American Needlework* (New York: Bonanza Books, 1938), chaps. 1 and 2.

27. Groves, *History of Needlework Tools and Accessories*, 19; see also Swabey, *Medieval Gentlewoman*, 81.

28. S. R. H. Jones, "The Development of Needle Manufacturing in the West Midlands before 1750," *Economic History Review*, n.s., 31, no. 3 (1978): 354–68. The making of needles, thimbles, and medieval cutlery in medieval Britain must have been small industries indeed, as they are not mentioned at all in L. F. Salzman, *English Industries of the Middle Ages*, 2nd ed. (Oxford: Clarendon Press, 1923), although there are twelve pages on the manufacture of bells (144–56).

29. Domenico Sella, "Peasants as Consumers of Manufactured Goods in Italy around 1600," in *European Peasant Family and Society*, ed. R. L. Rudolph (Liverpool: Liverpool University Press, 1995), 157, 163; and his *Crisis and Continuity: The Economy of Spanish Lombardy in the Seventeenth Century* (Cambridge: Harvard University Press, 1979), 16–17, 113.

30. William Stevenson and John Bridges, *A Ryght Pithy, Pleasaunt and Merie Comedie: Intytuled "Gammer Gurtons Needle"* (London: Thomas Colwell, 1575). There is a detailed discussion of this play in Wendy Wall, "'Household Stuff': The Sexual Politics of Domesticity and the Advent of English Comedy," *ELH: English Literary History* 65, no. 1 (1998): 1–45. On the storage of pins and needles, see Averil Colby, *Pincushions* (London: B. T. Batsford, 1975), 8.

31. Candace Wheeler, *The Development of Embroidery in America* (New York: Harper & Brothers, 1921), 103–4; and Julia Peterkin, "Black April," in *Quilt Stories*, ed. C. Macheski (1927; Lexington: University Press of Kentucky, 1994), 162.

32. John Stow and William Fitzstephen, *A Suruay of London* (London: John Windet for John Wolfe, 1598); see also "The Negro and the Needle," *Friends' Intelligencer* 13, no. 10 (1856): 155.

33. Groves, *History of Needlework Tools and Accessories*, 20. See also Hope Hanley, *Needlepoint in America* (New York: Scribner, 1969), 28.

34. Jones, "Development of Needle Manufacturing in the West Midlands," 354–55.

35. Christopher Dyer, *Everyday Life in Medieval England* (London: Hambledon Press,

1993), 263, 276; Ward, *English Noblewomen*, 61; and Woolgar, *Great Household*, 129. In remote areas, the tradition of buying needlework supplies with egg money persisted into the nineteenth century; there is a fictional reference in Frederick Marryat, *The Children of the New Forest* (1847; Oxford: Oxford University Press, 1991), 155.

36. On thimbles, see John W. E. Conybeare, *Roman Britain* (London: Society for Promoting Christian Knowledge, 1903), 177; on scissors, Katharine M. McClinton, *The Complete Book of Small Antiques Collecting* (New York: Coward–McCann, 1965), 202–4, and Joseph B. Himsworth, *The Story of Cutlery* (London: Benn, 1953), 151–57.

37. See, for example, "Handwork and Sewing: Fettered Cock Pewters," www.fettered cockpewters.com/page_sewing.htm (accessed 10 Dec. 2005).

38. Mary Eirwen Jones asserts in *A History of Western Embroidery*, 9, that embroidery hoops were used in the Middle Ages in Britain, but as her work has neither footnotes nor text references, her claim is difficult to verify. There is a brief bibliography on p. 156. According to Hope Hanley, embroidery frames "were never mentioned as being for sale" even in eighteenth-century America, suggesting that even at that late date, these tools were made to order. Hanley, *Needlepoint in America*, 46.

39. Kate Mertes, *The English Noble Household, 1250 to 1600* (Oxford, UK: B. Blackwell, 1988), 12.

40. Woolgar, *Great Household*, 50, 94, 100, 172, 186; Swabey, *Medieval Gentlewoman*, 95.

41. Urban T. Holmes, *Daily Living in the Twelfth Century* (Madison: University of Wisconsin Press, 1952), 83.

42. Laura Talbot, *The Gentlewomen* (1952; London: Virago, 1985), 74–75; and Arnold Bennett, *Imperial Palace* (Garden City, NY: Doubleday Doran, 1931), 263. The "gentlewoman" in Talbot's novel is offended by the necessity of doing plain sewing, which "seemed an insult to the intellect somehow" (p. 75).

43. On the consumption of candles in the late Middle Ages, see Woolgar, *Great Household*, 88–89. On the significance of lighting for fancywork, see also Swan, *Plain and Fancy*, 124.

44. Denis Diderot and Jean Le Rond d'Alembert, *Diderot Pictorial Encyclopedia of Trades and Industry* (New York: Dover, 1959), vol. 2, pl. 446.

45. Staniland, *Embroiderers*, 13.

46. Ward, *English Noblewomen*, 73.

47. Donald King et al., *The Victoria and Albert Museum's Textile Collection: Embroidery in Britain from 1200 to 1750* (London: The Museum, 1992), 16.

48. Iris Origo, *The Merchant of Prato* (New York: Knopf, 1957), 17.

49. Margaret Spufford, *The Great Reclothing of Rural England* (London: Hambledon Press, 1984), 21, 62, 64–66, 92, 94–98. On falconry gear, see Grassby, "Decline of Falconry," 45.

50. Woolgar mentions that medieval cookbooks were "not instructional manuals, with quantities and cooking times, but more of the nature of *aides-mémoire* to someone who might already know what they were doing." *Great Household*, 137.

51. Bert S. Hall, "Technical Treatises 1400–1600: Implications of Early Nonverbal Thought for Technologists; Abstract," *Technology and Culture* 19, no. 3 (1978): 485.

52. Meda P. Johnston and Glen Kaufman, *Design on Fabrics* (New York: Reinhold, 1967), 14–15.

53. Margaret H. Swain, *The Needlework of Mary, Queen of Scots* (New York: Van Nostrand Reinhold, 1973), 118–19. Swain probably means Konrad Gesner's 1560 *Icones Avium Omnium*, rather than the *Animalium* of the same year, which includes only viviparous species.

54. Stuart Robinson, *A History of Printed Textiles* (Cambridge: MIT Press, 1969), 11–12.

55. Cyril Bunt, "An Embroidery Pattern-Book," in *Needlework: An Historical Survey*, ed. Betty Ring (New York: Universe Books, 1975), 16. On the challenges of human figures for hobby needleworkers, see Mildred J. Davis, *Embroidery Designs, 1780–1820* (New York: Crown Publishers, 1971), xi; King et al., *Victoria and Albert Museum's Textile Collection*, 1:17–19; Harold Donaldson Eberlein, Abbot McClure, and Mabel F. Bainbridge, *The Practical Book of Early American Arts and Crafts* (Philadelphia: Lippincott, 1916), 97.

56. Cennino Cennini, *The Craftsman's Handbook*, ed. Daniel Thompson (New York: Dover, 1954), 105.

57. Swain, *Needlework of Mary, Queen of Scots*, 50–51.

58. Paul L. Grigaut, *Decorative Arts of the Italian Renaissance, 1400-1600* (Detroit: Institute of Arts, 1958), 75–76.

59. Marie-Anne Privat-Savigny, *Quand les Princesses d'Europe Brodaient* (Paris: Réunion des Musées Nationaux, 2003), 16–24.

60. Arthur Lötz, *Bibliographie der Modelbücher*, 2nd ed. (Stuttgart: A. Hiersemann, 1963). See also Averil Colby, *Samplers* (London: B. T. Batsford, 1964), 25.

61. Mary E. Jones, for example, mentions Katherine of Aragon as a lace hobbyist (*A History of Western Embroidery*, 32). Sylvia Groves mentions cutwork lace, actually an embroidery technique, as a significant source of demand for needlework cutlery in the late sixteenth century (*History of Needlework Tools and Accessories*, 45).

62. Federico de Vinciolo, *Les Singuliers et Nouveaux Pourtraicts* (Paris: Jean le Clerc, 1587). On other Italian designers of needlework, see Jones, *History of Western Embroidery*, 119–20. The development of pattern books of this type was significant to weaving as well as lace and embroidery; see S. A. Papadopoulos, *Greek Handicraft* (Athens: National Bank of Greece, 1969), 190.

63. Marcus Huish, *Samplers and Tapestry Embroideries*, 2nd ed. (New York: Dover, 1970), pl. VI and figs. 16 and 42.

64. Margaret B. Schiffer, *Historical Needlework of Pennsylvania* (New York: Scribner, 1968), 18.

65. King et al., *Victoria and Albert Museum's Textile Collection*, 18 and fig. 48.

66. Staniland, *Embroiderers*, 62.

67. Colby, *Samplers*, 137–42, 195. See also Huish, *Samplers and Tapestry Embroideries*, 110, where he reports the view of his fellow collector and connoisseur Mrs. C. J. Longman that darning samplers may originally have been a German innovation. Hope Hanley, however, opines in *Needlepoint in America* that "darning samplers are almost always English" (33).

68. Colby, *Samplers*, 147. Colby cites Donald King's 1960 *Samplers* as her source for this quote; the original text is the last part of Riche's eight-part *Farewell to Military Profession*. Riche is best known as the source of some of Shakespeare's plots and themes.

69. Richard Brathwait, *The English Gentlewoman* (1631), 225, quoted in John Taylor, Kathleen Epstein, and James Boler, *The Praise of the Needle: John Taylor's Poem of 1631* (Austin: Curious Works Press, 1995), 50.

70. On this point, see Swan, *Plain and Fancy*, 204; and Huish, *Samplers and Tapestry Embroideries*, 115–16.

71. To mention only a handful of thousands of examples of these manuals: on cooking, John Murrell, *Murrel's Two Books of Cookerie and Carving*, 5th ed. (London: M. Flesher for John Marriot, 1638); on sex, Aristotle, *Aristotle's Master-Piece* (London: Printed for W.B., 1694); on farriery, Henry Bracken, *Farriery Improved* (London: printed for J. Clarke, 1737); on falconry, Juliana Berners, *This Present Boke Shewyth the Manere of Hawkynge and Hunt-ynge* (Westmestre: Wynkyn the Worde, 1496); and on gardening, Thomas Hill, *A Most Briefe and Pleasaunt Treatyse, Teachynge Howe to Dress, Sowe, and Set a Garden* (London: Thomas Marshe, 1563).

72. Giovanni B. Ciotti, *A Booke of Curious and Strange Inventions, Called the First Part of Needleworkes* (London: J. Danter for William Barley, 1596).

73. Peter Burke, "The Invention of Leisure in Early Modern Europe," *Past and Present* 146 (1995): 136–50.

74. Jefferson to Charles W. Peale, 20 Aug. 1811, in *The Writings of Thomas Jefferson*, ed. Andrew A. Lipscomb and Albert Ellery Bergh (Washington, DC: Thomas Jefferson Memorial Association of the United States, 1903–4), 13:79.

75. There is an illuminating account of this kind of local knowledge in James C. Scott, *Seeing Like a State* (New Haven: Yale University Press, 1998), 333–39.

76. See, for example, M. Grieve, *A Modern Herbal* (New York: Dover, 1971), 1:354–57, on the challenges of cultivating ginseng.

77. William Eamon, *Science and the Secrets of Nature* (Princeton: Princeton University Press, 1994), 7. For a discussion of these manuals in popular culture, see chap. 7 of Eamon's work (234–66).

78. See, for example, Marianne M. C. Alford, and Elizabeth G. Holahan, *Needlework as Art* (London: S. Low, Marston, Searle, and Rivington, 1886).

79. On this point, see Therle Hughes, *English Domestic Needlework, 1660–1860* (New York: Macmillan, 1961), 227–36; and Schiffer, *Historical Needlework of Pennsylvania*, 26.

80. For a definition of bibliometrics, see the Web page by information scientist Ruth Palmquist, "Bibliometrics," University of Texas at Austin, Graduate School of Library and Information Science, www.gslis.utexas.edu/~palmquis/courses/biblio.html (accessed 2 Jan, 2006). Palmquist defines bibliometrics as a research method that "utilizes quantitative analysis and statistics to describe patterns of publication within a given field or body of literature." Information scientists typically use it in more complex and nuanced ways— analysis of citations across journal titles over time, for example—than the simple counts of titles I employ here. The bibliometric calculation of what is known as the "journal impact factor (JIF)," for example, has become so significant to scientific publishing that there are rumors of manipulation of citations in journal articles to artificially skew the JIF. See Richard Monastersky, "The Number That's Devouring Science," *Chronicle of Higher Education*, 14 Oct. 2005, http://chronicle.com/weekly/v52/i08/08a01201.htm.

81. Some historians are well ahead of me in the use of bibliometrics to illuminate the

development of a technical literature. See, for example, Miriam G. Reumann, *American Sexual Character* (Berkeley: University of California Press, 2005), 6.

82. *Make and Mend for Victory* (New York: Spool Cotton Co., 1942).

83. Wayne Muller and Stuart Muller, *Darn it!* (Gas City, IN: L–W Book Sales, 1995). See also darners in the collection of Colonial Williamsburg, Abby Aldrich Rockefeller Folk Art Museum, accession no. G1971-2051.

84. M. W., *A Queen's Delight* (London: Nathaniel Brook, 1655).

85. M. W., *The Queen's Closet Opened* (London: Nathaniel Brooke and C. Eccleston, 1662).

86. Juan L. Vives, *A Very Frutefull and Pleasant Boke Called the Instruction of a Christen Woman*, trans. R. Hyrde (London: Thomas Berthelet, 1529), bk. I, chap. 3; see also Ruth Kelso, *Doctrine for the Lady of the Renaissance* (Urbana: University of Illinois Press, 1956), 46, 112.

87. Hannah Woolley, *The Accomplish'd Lady's Delight in Preserving, Physick, Beautifying, and Cookery* (London: B. Harris, 1675); and Elizabeth Kent, *A True Gentlewoman's Delight* (London: printed by G.D. for W.I., 1653). The British title by "P. T." *Accomplish'd Lady's Delight in Preserving, Physick, Beautifying, Cookery and Gardening* (London: Benjamin Harris), in its ninth edition by 1706, offered recipes for almond caudle, candied borage flowers, quince chips, and cinnamon water, among other elegant delicacies (3, 23, 34, 47).

88. Adeline Cole presents examples from both literatures in her brief essay, "The 17th-Century Still-Room," in Daniel Foley, *Herbs for Use and for Delight* (New York: Dover, 1974), 274–75.

89. Hannah Woolley, *The Compleat Servant-Maid*, 4th ed. (London: Thomas Passinger, 1685). That this is a fourth edition indicates it must have been a popular work.

90. Sanjida O'Connell, *Sugar: The Grass That Changed the World* (London: Virgin Books, 2004). On the human cost of this sugar, see Hugh Thomas, *The Slave Trade* (New York: Simon & Schuster, 1997).

91. For examples, see Woolley, *Compleat Servant-Maid*; John Partridge, *The Treasurie of Commodious Conceits* (London: Richarde Jones, 1573); Thomas Dawson, *The Second Part of the Good Hus-Wives Jewell* (London: E. Allde for Edward White, 1597); *A Closet for Ladies and Gentlewomen* (London: F. Kingston for Arthur Johnson, 1608); John Murrell, *A Daily Exercise for Ladies and Gentlewomen* (London: T. Snodham for the Widow Helme, 1617); *A Book of Fruits and Flowers* (London: M.S. for Thomas Jenner, 1653); Elizabeth Kent, *A Choice Manual of Rare and Select Secrets in Physick and Chirurgery* (London: R. Norton, 1653); and Nicolas de Bonnefons, *Le Jardinier François, qui Enseigne à Cultiver les Arbres, and Herbes Potagères,* 8th ed. (Paris: Chez A. Cellier, 1666).

92. Susan G. Bell, "Women Create Gardens in Male Landscapes: A Revisionist Approach to Eighteenth-Century English Garden History," *Feminist Studies* 16, no. 3 (1990): 471–91. There is a fictional portrayal of the role of hobby gardening among the English aristocracy in Philippa Gregory's novels of the English Civil War, *Earthly Joys* and *Virgin Earth* (New York: Touchstone, 2005 and 2006).

93. Francis Drope, *A Short and Sure Guid in the Practice of Raising and Ordering of Fruit-Trees* (Oxford: Ric. Davis, 1672); see also John Laurence, *The Clergy-Man's Recreation; Shewing the Pleasure and Profit of the Art of Gardening,* 5th ed. (London: Printed for B. Lintott, 1717).

94. Leonard Meager, *The English Gardner* (London: J. Rawlins for M. Wotton and G. Conyers, 1688); and Michael MacCaskey, *Gardening for Dummies* (Foster City, CA: IDG Books, 1996).

95. John Parkinson, *Paradisi in Sole Paradisus Terrestris*, 2nd ed. (London: R.N. for Richard Thrale, 1656); William Hughes, *The Flower-Garden* (London: H.B. for William Crook, 1671); and Samuel Gilbert, *Florist's Vade Mecum*, 2nd ed. (London: Thomas Simmons, 1683). In 1681, according to the *OED*, a "florist" was a flower gardener rather than a retailer of flowers and greenery.

96. See, for example, Rex Stout, "Zero Clue," in *Three Men Out: A Nero Wolfe Threesome* (New York: Bantam Books, 1955), 53–103. It is clear that for Wolfe himself (although not for his professional cook, Fritz), cooking is also a hobby. On pineapples in Scotland, see Sir James Justice, *The Scots Gardiners' Director*, 2nd ed. (Edinburgh: A. Donaldson, 1759); on oranges in Holland see Henrik van Oosten, *De Nederlandschen Hof* (Rotterdam: Dirk Vis, 1784). Orange growing was attempted in Germany as well; see Johann D. Fülck, *Neue Garten Lust* (Augsburg: Johann Andreas Pfeffel, 1720).

97. Thomas Barker, *The Country-Man's Recreation* (London: T. Mabb, for William Shears, 1654).

98. Matthew Stevenson, *The Twelve Moneths* (London: M.S. for T. Jenner, 1661).

99. S. J., *Profit and Pleasure United* (London: S. Lee and Andrew Thorncome, 1684). The quotation in the text is actually what appears on the title page.

100. Joseph Blagrave, *The Epitome of the Art of Husbandry*, 2nd ed. (London: Benjamin Billingsley, 1670).

101. See, for examples, Gervase Markham, *Country Contentments*, 6th ed. (London: William Wilson for John Harison, 1649); H.R., *The School of Recreation* (London: H. Rodes, 1684).

102. John White, *The Art of Ringing* (Little-Britain: G. Conyers, 1680).

103. Nathaniel B. Nye, *The Art of Gunnery* (London: William Leak, 1670).

104. Ruth S. Cowan, *More Work for Mother* (New York: Basic Books, 1983), 26.

105. The guide included a section on how "to heal all diseases in men or women with chew'd white bread," this last no doubt a godsend after overindulgence in home brew. Gervase Markham, *The Compleat Husbandman and Gentleman's Recreation* (London: G. Conyers, 1695). Samuel Child, *Every Man His Own Brewer* (London: Printed for the Author, 1790), had gone through six editions by 1790. See also "The Practical Brewer" [pseud.], *A Guide to Young Brewers: Particularly Adapted to the Families of the Nobility and Gentry, Farmers, and Private Brewers* (London: Printed for the Author, 1820); W. Brande, *The Town and Country Brewery* (London: Dean and Munday, 1830); and the anonymous *Complete Family Brewer* (Philadelphia: B. Graves, 1805).

106. George Wright, *The Bird-Fancier's Recreation* (London: T. Ward, 1728). On horses and dogs, see, for example, Nicholas Cox and John Manwood, *The Gentleman's Recreation in Four Parts*, 4th ed. (London: I. Dawks for D. Browne and N. Rolls, 1697); Gerard Langbaine, *The Hunter* (Oxford: L. Lichfield for Nicolas Cox, 1685); and John Caius and Abraham Fleming, *Of Englishe Dogges* (London: John Charlewood for Rychard Johnes, 1576). The original edition of this last work was published in 1570 in Latin, but the book was translated into the vernacular only six years later, suggesting a demand from those lacking the benefits of a classical sixteenth-century education.

107. Charles D. Mallary et al., *The Little Boy's Own Book, Consisting of Games and Pastimes* (London: H. Allman, Sears, 1850).

108. Keith R. Gilbert, *Textile Machinery* (London: Science Museum, 1971).

109. For key elements of this process, see David J. Jeremy, *Transatlantic Industrial Revolution* (North Andover, MA: Merrimack Valley Textile Museum; Cambridge, MA: MIT Press, 1981).

Chapter 3 · Hedonization and Industrialization

1. Nerylla D. Taunton, *Antique Needlework Tools and Embroideries* (Woodbridge, Suffolk, UK: Antique Collectors' Club, 1997), 15. Like all book-length histories of needlework tools except those of sewing machines, Taunton's work was written from the perspective of an antique collector.

2. "Tatting Continues to Gain Popularity," *Star Needlework Journal* 5, no. 1 (1920): 4; *Richardson's Tatting Book*, bk. 7 (Chicago: Richardson Silk Co., 1900); Emma Farnes, *Baby's and Child's Art Needle work*, bk. 1 (St. Louis: E. C. Spuehler, 1900); Mary E. Fitch, *Crocheted Yokes and Tatting* (Brookline, MA: Fitch, 1915); and Julia Sanders, *The Priscilla Tatting Book*, bk. 2 (Boston, MA: Priscilla Publishing Co., 1915).

3. Marion L. Channing, *Textile Tools of Colonial Homes* (Marion, MA: The Author, 1969), 53; and Katharine M. McClinton, *The Complete Book of Small Antiques Collecting* (New York: Coward-McCann, 1965), 205.

4. Hope Hanley, *Needlepoint in America* (New York: Scribner, 1969), 34–64.

5. Mildred J. Davis, *Early American Embroidery Designs* (New York: Crown Publishers, 1969), 25.

6. Rita H. Gottesman, *The Arts and Crafts in New York, 1800–1804* (New York: New-York Historical Society, 1965), 95–109. See also advertisements for embroidery pattern drawing in the first volume of this series: Gottesman, *The Arts and Crafts in New York, 1726–1776* (New York: New-York Historical Society, 1938), 275–77. On fishhooks see Mary Landon and Susan Swan, *American Crewelwork* (New York: Macmillan, 1970), 31–32.

7. Colonial Williamsburg's Abby Aldrich Rockefeller Folk Art Museum has a number of examples of these luxurious needlework tools and containers, including accession numbers G1971-3485 and G1971-3459. The collection also includes a twentieth-century set of leather-handled needlework tools made by Tiffany & Co., G1971-3482.

8. Harold D. Eberlein, Abbot McClure, and Mabel F. Bainbridge, *The Practical Book of Early American Arts and Crafts* (Philadelphia: Lippincott, 1916), 97. On the difficulties of realism in painting and needlework, see Jean Lipmann and Alice Winchester, *The Flowering of American Folk Art, 1776–1876* (New York: Viking Press in cooperation with the Whitney Museum of American Art, 1974), 50.

9. George F. Dow, *The Arts and Crafts in New England, 1704–1775* (New York: Da Capo Press, 1967), 261, 267–68, 274–76.

10. Margaret B. Schiffer, *Historical Needlework of Pennsylvania* (New York: Scribner, 1968), 17, 100, 111; Susan Burroughs Swan, *Plain and Fancy* (New York: Holt, Rinehart and Winston, 1977), 96.

11. See, for example, *The Lady's Magazine* (London: Robinson and Roberts, 1776).

12. Elizabeth S. Drinker, *The Diary of Elizabeth Drinker*, ed. Elaine F. Crane (Boston: Northeastern University Press, 1991). For a few examples of many in this diary, see, in vol. 1, pp. 1–3 (1758); 18, 34, 38 (1759), 47, 50–51, 59–61, 65, 68, 70, 76 (1760); 87, 106 (1761); and 277 (1778); and in vol. 3, p. 1777 (1804).

13. Sam B. Warner, *The Private City* (Philadelphia: University of Pennsylvania Press, 1968), 18.

14. Mildred J. Davis, *Early American Embroidery Designs,1780–1820* (New York: Crown, 1971), 25. On reading, see the Patty Polk sampler of 1800 (ibid., 23); see also Drinker, *Diary*, 3:1791 (1804) and 1995 (1806). Card-playing and other forms of genteel gaming were considered acceptable for ladies in Britain, according to "G.R.," *The Accomplish'd Female Instructor* (London: Knapton, 1704), 35–36, but Jefferson disapproved of both for American women. On the advantages of needlework as a camouflage for eavesdropping on eighteenth-century male conversation, see Charles B. Brown, *Wieland* (1798; Oxford: Oxford University Press, 1994), 28.

15. Hannah More, "Lady Pennington's Unfortunate Mother's Advice to Her Daughters," in *The Lady's Pocket Library* (Philadelphia: Mathew Carey, 1792), 146, quoted in Swan, *Plain and Fancy*, 73.

16. Sarah S. and Sarah A. Emery, *Reminiscences of a Nonagenarian* (Newburyport, MA: W. H. Hues Printers, 1879), 21.

17. Anne H. S. Livingston and Ethel Arms, *Nancy Shippen, Her Journal Book* (Philadelphia: J. B. Lippincott, 1935), 220–21, quoted in Swan, *Plain and Fancy*, 150.

18. *Godey's Lady's Book*, July 1830, quoted in Davis, *Early American Embroidery Patterns*, 28. This was in the period before Sarah Josepha Buell Hale (1788–1879) took over editorship of the magazine.

19. Swan, *Plain and Fancy*, 12, 25. Even Mary Wollstonecraft thought working-class girls should learn plain work; see her *Vindication of the Rights of Woman*, 2nd ed. (1796; New York: Norton, 1988), 75, 293, 295. There is a fictional reference to British Navy sailors darning their own socks in Ian McEwen, *Atonement* (New York: Anchor Books, 2003), 270; see also Paul Watkins, *The Story of My Disappearance* (New York: Picador, 1998), 49; and the same author's *The Forger* (London: Faber, 2000), 115.

20. Photographs of darning samplers can be found on the Internet; for example, good photos of at least three darning samplers can be seen at http://needlenecessities.blogspot.com/2007/09/dutch-samplers-part-two-darning.html (accessed 13 Nov. 2008).

21. Schiffer, *Historical Needlework of Pennsylvania*, 58; Georgian Brown Harbeson, *American Needlework* (New York: Bonanza Books, 1938), 75–76. See also Ellen J. Gehret, *This Is the Way I Pass My Time* (Birdsboro, PA: Pennsylvania German Society, 1985), 9–80.

22. Marcus Huish, *Samplers and Tapestry Embroideries*, 2nd ed. (New York: Dover, 1970), 22.

23. Ibid., 122.

24. See, for two of many examples, *An Account of Charity-Schools in Great Britain and Ireland*, 11th ed. (London: Joseph Downing, 1712), 18; and *An Account of Several Work-Houses for Employing and Maintaining the Poor* (London: Joseph Downing, 1725) 7

25. Davis, *Early American Embroidery Designs*, 32.

26. Schiffer, *Historical Needlework of Pennsylvania*, 27, 44, 53, 83, 106.

27. Harbeson, *American Needlework*, 53–54.

28. Drinker, *Diaries*, 1:xv.

29. Ibid., xviii; 18, 34, 38 (1759); 45–47, 50, 59–61, 65–66, 68, 70, 76 (1760); 87 (1761); 106 (1763); 277, 290, 304 (1778); 371 (1780); 461, 464 (1791); 537 (1793); 591, 603, 613, 617, 629 (1794); 642, 694–95, 702, 708, 712, 716, 733, 737, 746 (1795).

30. Therle Hughes, *English Domestic Needlework* (New York: MacMillan, 1961), 239.

31. See, on this point, Sylvia Groves, *History of Needlework Tools and Accessories*, Country Life Books (Feltham, UK: Hamlyn Publishing Group, 1966),100; and Schiffer, *Historical Needlework of Pennsylvania*, 59–60.

32. Lis Paludan, *Crochet* (Loveland, CO: Interweave Press, 1995), 10–22.

33. Pauline Turner, *Crochet* (Aylesbury, UK: Shire, 1984), 5, 9, 11, 20–21; see also Mary Konior, *Heritage Crochet* (London: Dryad Press, 1987).

34. William R. Bagnall, *Samuel Slater and the Early Development of the Cotton Manufacture in the United States* (Middletown, CT: n.p., 1890).

35. Charlotte Brontë, *Jane Eyre* (London: Smith, Elder, 1847). On ironing and mending, see chaps. 1, 7, and 13; on garment cutting and sewing, chaps. 5, 6, 11, 21, 22, 28–29. On Jane's ladylike accomplishments, see chaps. 6, 10, 17, and 32. For the brown Holland workpockets, see chap. 4. Mrs. Fairfax and her knitting appear in chaps. 11, 13, 14, 16, and 22 and Hannah's in chap. 28. Hannah's other duties are mentioned in chap. 29. The humble characters of plain sewing and knitting are stressed by Mr. St. John in chap. 30 and by Jane herself in chap. 31.

36. Charlotte M. Yonge, *The Pillars of the House* (Leipzig: B. Tauchnitz, 1873), 1:280. Wilmet's diligent attention to plain work is mentioned elsewhere in the novel as well, in 2:268 and 4:100, where she is contrasted with her younger sister Robina, who is making lace. Toward the close of the novel, Wilmet is finally permitted the leisure to tat (5:184).

37. Louisa M. Alcott, *Little Women* (1869; Ware, Hertfordshire, UK: Wordsworth Classics, 1993), 9–10, 15–16, 43, 62–63, 115, 135–42, 161, 207, 216; and Jane Austen, *Pride and Prejudice* (1813; Toronto: Bantam Books, 1981), 28, 34, 40, 216, 250, 267.

38. For examples, see William M. Thackeray, *The Newcomes* (1854; Ann Arbor: University of Michigan Press, 1996), 747; Margaret Oliphant and Thomas Bailey Aldrich, *The Second Son* (Boston: Houghton Mifflin, 1888), 72; and Oliphant's *Harry Joscelyn* (Leipzig: B. Tauchnitz, 1881), 1:169–70. See also a young woman's refusal to dissemble in Anthony Trollope, "Alice Dugdale," in *Complete Shorter Fiction*, ed. J. Thompson (New York: Carroll & Graf, 1992), 865; this story was originally published in *Harper's New Monthly*, Aug. 1860.

39. For examples of this convention regarding novel-reading, see Mrs. Henry Wood, *The Red Court Farm* (Leipzig: Tauchnitz, 1868), 1:75, 113, 134 and 2:44; Mary E. Braddon, *The Doctor's Wife* (Leipzig: B. Tauchnitz, 1864), 1:148–49, 157, 160, 273, and 2:166; Mrs. Alexander, *A Ward in Chancery* (Leipzig: B. Tauchnitz, 1894), 1:8, 48; Charlotte M. Yonge, *A Reputed Changeling* (Leipzig: B. Tauchnitz, 1890), 1:139, 164, 265–66; Margaret Oliphant, *The Minister's Wife* (London: Hurst and Blackett, 1869), 1:245 and 2:7, 39-40, 87, 279; and Oliphant, *The Heir Presumptive and the Heir Apparent* (Leipzig: B. Tauchnitz, 1892), 1:44; E. H. Young, *The Curate's Wife* (1934; New York: Penguin, 1985), 93, 180–81;

and Florence Marryat, *Petronel* (Leipzig: B. Tauchnitz, 1870), 1:93, 269–70. In W. E. Norris's 1897 antifeminist novel *Clarissa Furiosa* (Leipzig: Tauchnitz), the chief protagonist does no needlework of any kind until she repents of her efforts to foment gender revolution; see 2:290.

40. Wilkie Collins, "*I Say No*" (Leipzig: B. Tauchnitz, 1884), 2:79; and Anthony Trollope, *The Macdermots of Ballycloran* (1847; London: Penguin, 1993), 61. For an (as it were) novel view of fiction-reading as a primrose path to masturbation, see Thomas W. Laqueur, *Solitary Sex* (New York: Zone Books, 2003), 320–29. Needlework, while conducive to fantasy, at least kept the hands busy and in plain sight.

41. On this point, see Cowan, *More Work for Mother*, 157.

42. Eliza W. R. Farrar, *The Young Lady's Friend* (1836; New York: Arno, 1974), 124–25; and Alcott, *Little Women*: 135. Charlotte Brontë's character Mr. Brocklehurst, the parsimonious schoolmaster of Lowood in *Jane Eyre*, chap. 6, apparently shares Farrar's views on the importance of properly darned stockings.

43. On the gender of the one reading aloud, see Farrar, *Young Lady's Friend*, 123; on the supposedly elevated station of men, see, for example, p. 403. When the boy Laurie joins the March girls' "Busy Bee Society" in Alcott's *Little Women*, his role is that of reader to the knitters (141).

44. Dorothy Canfield Fischer, *The Home-Maker* (New York: Harcourt Brace, 1924), 198, 312.

45. In fact, darning and mending were often the responsibility of the laundress or laundry maid. See Hannah Woolley, *The Compleat Servant*, 7th ed. (London: E. Tracy, 1704), 125; see also Arwen Mohun, *Steam Laundries* (Baltimore: Johns Hopkins University Press, 1999), 93.

46. Arnold Bennett, *Imperial Palace* (Garden City, NY: Doubleday, Doran, 1931), 263; on boardinghouses, see E. H. Young, *Jenny Wren* (1932; New York: Penguin, 1985), 74.

47. Mary D. Russell, *A Thread of Grace* (New York: Random House, 2005), 40–46; and Louise Erdrich, *Four Souls* (New York: Harper Perennial, 2005), 176. An equally suspicious document, though purportedly nonfiction, is Margaret H. Carter, *Her Hobby Is Mending* (Burlington, VT: University of Vermont and State Agricultural College, 1947).

48. Mary R. Walker and Myra F. Eells, *On to Oregon*, ed. Clifford M. Drury (Lincoln: University of Nebraska Press, 1998), 300.

49. See, for example, the beautifully illustrated J. M. Holt, *The Art and Practice of Mending* (London: Sir I. Pitman & Sons, 1933).

50. Clarice L. Scott, *ABC's of Mending* (Washington, DC: U.S. Department of Agriculture, 1942); and *Make and Mend for Victory* (New York: Spool Cotton Co., 1942)

51. Ursula Hegi, *Stones from the River* (New York: Scribner, 1994), 339; and E. H. Young, *Chatterton Square* (1947; London: Virago, 1987), 318.

52. See, for instance, Bell I. Wiley, *The Life of Johnny Reb* (Indianapolis: Bobbs-Merrill, 1943), 116. One of these sewing kits is depicted in Florence H. Pettit, *America's Printed and Painted Fabrics, 1600–1900* (New York: Hastings House, 1970), 146; and they can be found in almost every historical society collection in the United States. Women also carried these, mainly as travel gear, or when leaving home for school. See Harbeson, *American Needlework*, 148–49. Fictional examples can be found in Sylvia Townsend Warner's missionary in

"Mr. Fortune's Maggot," in *Four in Hand* (1927; New York: Norton, 1986), 139; and Eden Phillpotts, *The Thief of Virtue* (London: Macmillan, 1927), 68.

53. Betty Miller, *On the Side of the Angels* (1945; London: Virago, 1985), 167.

54. See, for example, the highly developed early-nineteenth-century examples in Swan, *Plain and Fancy*, plate 20. Gold and silver laces were being professionally hand-knitted in very fine gauges by Moslem women in India as late as the 1950s; see Rustam J. Mehta, *The Handicrafts and Industrial Arts of India* (Bombay: Taraporevala, 1960), 105.

55. Richard Rutt, *History of Hand Knitting* (Loveland, CO: Interweave Press, 1987), 8–9, 28–29. Readers unfamiliar with the structure of knitting and other single-element techniques should consult Irene Emery's illuminating and authoritative handbook, *The Primary Structures of Fabrics* (Washington, DC: The Textile Museum, 1966), 29–49. This chapter also clarifies and illustrates the distinction between knitting and crochet.

56. Rudolf P. Hommel, *China at Work* (New York: John Day Co., for the Bucks County Historical Society of Doylestown, PA, 1937), 208.

57. Virginia S. Gifford, *An Annotated Bibliography on Hand Knitting* (Washington DC: The Author, 1969), 3.

58. Ibid., 4; and Irena Turnau, "The Diffusion of Knitting in Medieval Europe," in *Cloth and Clothing in Medieval Europe* (Exeter, NH: Heinemann Educational Books, 1983).

59. Margaret Oliphant notes this reduced demand for lighting in her novel, *It Was a Lover and His Lass* (Leipzig: Tauchnitz, 1883), 2:59. On this point, see also Swan, *Plain and Fancy*, 25.

60. See, for example, Farrar, *The Young Lady's Friend*, 25; and Alcott, *Little Women*.

61. Emery, *Primary Structures of Fabrics*.

62. Sonna Noble and Theodore H. Levin, *How to Successfully Operate a Knitting Shop* (Washington, DC: Public Affairs Press, 1961), 88–94.

63. On knitting as a path to spiritual enlightenment, see recent titles like Bernadette Murphy, *Zen and the Art of Knitting* (Avon, MA: Adams Media Corp., 2002); Susan S. Izard and Susan S. Jorgensen, *Knitting into the Mystery* (Harrisburg, PA: Morehouse Publishing, 2003); and Tara J. Manning, *Mindful Knitting* (Boston: Tuttle Publishing, 2004)

64. Claudia B. Kidwell, *Cutting a Fashionable Fit* (Washington, DC: Smithsonian Institution Press, 1979), 18.

65. Carole Shammas, *The Pre-Industrial Consumer in England and America* (Oxford: Clarendon, 1990), 65; Frederick A. Wells, *The British Hosiery and Knitwear Industry* (New York: Barnes & Noble, 1972), 15–23; Clifford Gulvin, *The Scottish Hosiery and Knitwear Industry, 1680–1980* (Edinburgh: J. Donald, 1984), 1–64; Lynn Abrams, *Myth and Materiality in a Woman's World: Shetland, 1800–2000* (Manchester: Manchester University Press, 2005), 37, 57–58, 65, 77, 84, 98–119, 141, 200; Linda G. Fryer, "Knitting by the Fireside and on the Hillside," *A History of the Shetland Hand Knitting Industry, c. 1600–1900* (Lerwick: Shetland Times, 1995), 4–32; Averil Colby, *Samplers* (London: B. T. Batsford, 1964), 200; and Joan Thirsk, "The Fantastical Folly of Fashion," in *Textile History and Economic History: Essays in Honour of Miss Julia de Lacy Mann*, ed. N. B. Harte and K. G. Ponting (Manchester: Manchester University Press, 1973).

66. Charlotte M. Yonge, *Dynevor Terrace* (Leipzig: B. Tauchnitz, 1857), 1:44.

67. Tom Whayne, Michael E. DeBakey and John B. Coates, *Cold Injury, Ground Type*

(Washington, DC: Office of the Surgeon General, 1958) chap. 1, esp. p. 7; and Erna Risch and Thomas M. Pitkin, *Clothing the Soldier of World War II* (Washington, DC: Government Printing Office, 1946), 97–100.

68. For a fuller discussion of how gloves are hand-knit to measure, see Mary Thomas, *Mary Thomas's Knitting Book* (New York: Morrow, 1938), 197–210.

69. See, for example, Janet Rehfeldt and Mary Jane Wood, *Crocheted Socks!* (Woodinville, WA: Martingale, 2003). For discussions of the importance of knitted socks as military footwear other than those in note 67, above, see Peter R. Mansoor, *The GI Offensive in Europe* (Lawrence, KS: University Press of Kansas, 1999), 201, 230–31; and Edna Goodheart, "Comforts for Our Soldiers," *Needlecraft* 9, no. 4 (December 1917): 6.

70. On the various methods of turning sock heels and shaping toes, see Thérèse de Dillmont, *Encyclopedia of Needlework*, 2nd ed. (1884; Philadelphia: Running Press, 1978), 231–38; and Thomas, *Mary Thomas's Knitting Book*, 211–32. On home knitting of hosiery, see Alice Hyneman Rhine, "Woman in Industry," in *Woman's Work in America*, ed. A. N. Meyer (1891; New York: Arno Press, 1972).

71. William Felkin, *Felkin's History of the Machine-Wrought Hosiery and Lace Manufactures* (Newton Abbot, Devon: David & Charles, 1967); and Milton N. Grass, *History of Hosiery* (New York: Fairchild, 1955).

72. Marie Hartley and Joan Ingilby, *The Old Hand-Knitters of the Dales*, 2nd ed. (Clapham: Dalesman, 1969). See also "History of the Dales," www.dalenet.co.uk/features/knitting/history5.htm (accessed 9 March 2006).

73. Europeans find it irritating to watch Americans and British knitters needlessly (no pun intended) yarning over with every stitch. For a description of this difference, see Thomas, *Mary Thomas's Knitting Book*, 46–48.

74. Alcott, *Little Women*, 142.

75. Thomas, *Mary Thomas's Knitting Book*, 6–10; see also Gifford, *Annotated Bibliography*, 10.

76. Harriet Beecher Stowe, "The Minister's Wooing," in *Quilt Stories*, ed. C. Macheski (1859; Lexington: University Press of Kentucky, 1994), 81.

77. See, for example, the fictional example of a lovelorn knitter in Anthony Trollope, "The Two Heroines of Plumplington," in *Complete Shorter Fiction*, ed. J. Thompson (1860; New York: Carroll & Graf Publishers, 1992), 932. Sarah Lyddon Morrison asserts that such knitting jinxes relationships; see her *Modern Witch's Spellbook* (New York: David McKay, 1971), 33. On knitting and other needlework for babies, see the fictionalized account in Elizabeth Taylor, *A Game of Hide and Seek* (1951; London: Virago Press, 1986), 60, 112.

78. Knitting needles are referred to as "wires" in Margaret Oliphant, *Kirsteen* (Leipzig: B. Tauchnitz, 1891), 1: 13.

79. Thomas, *Mary Thomas's Knitting Book*. The collections of Colonial Williamsburg's Abby Aldrich Rockefeller Folk Art Museum contain examples of both industrially produced and artisan-made knitting needles (accession numbers 1964.406.15, G1971.1253.1-13; and G1971.3465.1-17). See also Laurel T. Ulrich, *The Age of Homespun* (New York: Knopf, 2001), 374–412.

80. See, for example, the knitting needles in T. M. James, *Longmans' Complete Course of Needlework, Knitting, and Cutting Out* (London: Longmans, Green, 1901), 331.

81. Swan, *Plain and Fancy*, 132, 160. On the origins of the art, see Elizabeth Mincoff and Margaret S. Marriage, *Pillow Lace* (New York: Dutton, 1907), 9–56, as well as the two connoisseurs' guides: Emily Jackson and Ernesto Jesurum, *A History of Hand-Made Lace* (1900; repr., Detroit: Gale Research 1971); and T. L. Huetson, *Lace and Bobbins* (South Brunswick, NJ: A. S. Barnes, 1973).

82. Charlotte Kellogg, *Bobbins of Belgium* (New York: Funk & Wagnalls, 1920); Emily Noyes Vanderpoel and Elizabeth C. Barney Buel, *American Lace and Lace-Makers* (New Haven: Yale University Press, 1924); and Katrina Honeyman, "The Lace Industry: Organization and Finance," in *Origins of Enterprise* (Manchester, UK: Manchester University Press; New York: St. Martin's Press, 1983), 115–27. There is a vivid fictional reference to the poverty of French lacemakers in the late seventeenth century in Yonge, *Reputed Changeling*, 2:52.

83. Anne Buck, *Thomas Lester, His Lace and the East Midlands Industry, 1820–1905* (Bedford, UK: Ruth Bean, 1981); and Huetson, *Lace and Bobbins*, 77.

84. See, for a few of many examples, Mary M. Atwater, *The Shuttle-Craft Book of American Hand-Weaving* (New York: Macmillan, 1928); Edward F. Worst, *Foot-Power Loom Weaving*, 3rd ed. (Milwaukee: Bruce Publishing Co., 1924); Thomas Woodhouse, *The Handicraft Art of Weaving* (London: H. Frowde, 1921); and Luther Hooper, *Hand-Loom Weaving* (London: Sir I. Pitman & Sons, 1936). For a discussion of the pleasures of handweaving, see Anni Albers, *On Weaving* (Middletown, CT: Wesleyan University Press, 1965).

85. Louisa A. Tebbs and Rosa Tebbs, *The Art of Bobbin Lace*, 3rd ed. (London: Chapman & Hall, 1911); and Margaret Maidment, *A Manual of Hand-Made Bobbin Lace Work* (London: Pitman, 1931).

86. Osma G. Tod, *Bobbin Lace* (Coral Gables, FL: Tod Studio, 1969); and Kaethe Kliot and Jules Kliot, *Bobbin Lace* (New York: Crown, 1973).

87. Kaethe Kliot died in 2002; Jules Kliot is still associated with the Lacis Museum in California, which houses the Kliot collection of textile arts. See Zelda Bronstein, "The Art That Saved the Irish from Starvation," *Berkeley Daily Planet*, 19 April 2005, online at www .berkeleydaily.org/article.cfm?issue=04-19-05&storyID=21207 (accessed 1 June 2006).

88. Members of this organization periodically complain that the name suggests either tight corsetry, surreptitious alcohol consumption, or both, but member plebiscites on changing the name have so far all gone down to defeat. See www.internationaloldlacers .org/index.html (accessed 1 June 2006) for an overview of the organization and its history.

89. Ruth S. Cowan, *More Work for Mother* (New York: Basic Books, 1983), 65; Swan, *Plain and Fancy*, 23; Shammas, *The Pre-Industrial Consumer in England and America*, 200; and Catherine Fennelly, *Textiles in New England, 1790–1840* (Sturbridge, MA: Old Sturbridge Village, 1961), 37.

90. Ivy Pinchbeck, *Women Workers and the Industrial Revolution, 1750–1850* (London: G. Routledge, 1930), 287–90.

91. J. H. Ingraham, "The Milliner's Apprentice," *Godey's Lady's Book* 22 (1841): 194; and "Results of the Invention of the Sewing Machine," *Littel's Living Age* 134 (1877): 187–90.

92. Claudia B. Kidwell and Margaret Christman, *Suiting Everyone* (Washington, DC: Smithsonian Institution, 1974), 45; Dolores Janiewski, "Archives, Making Common

Cause: The Needlewomen of New York, 1831–69," *Signs* 1, no. 3 (1976): 777–86; and Joel I. Seidman, *The Needle Trades* (New York: Farrar & Rinehart, 1942), 80–81. By 1910, however, a U.S. Bureau of Labor report on women wage earners noted a decline in the proportion (though not the absolute numbers) of women compared with men employed in the needle trades. See Charles P. Neill, *Report on Condition of Woman and Child Wage-Earners in the United States* (Washington: United States Bureau of Labor, 1910), 488.

93. Egal Feldman, *Fit for Men* (Washington, DC: Public Affairs Press, 1960), 2–48; and Virginia Penny, *The Employments of Women* (Boston: Walker, Wise & Co., 1863), 110–14.

94. Mabel H. Willett, *The Employment of Women in the Clothing Trade* (New York: Columbia University Press, 1902), 31–46; and Margaret Walsh, "Industrial Opportunity on the Urban Frontier: 'Rags to Riches' and Milwaukee Clothing Manufacturers, 1840–1880," *Wisconsin Magazine of History* 57, no. 3 (1974): 174–94.

95. Massachusetts Bureau of Statistics of Labor, *The Working Girls of Boston*, from *The Fifteenth Annual Report of the Massachusetts Bureau of Statistics of Labor, for 1884* (Boston: Wright & Potter Printing Co., 1889), 103; Rosalyn Baxandall, Linda Gordon, and Susan Reverby, "Archives: Boston Working Women Protest, 1869" *Signs* 1, no. 3, pt. 1 (1976): 803–8; Jo Anne Preston, "'To Learn Me the Whole of the Trade': Conflict between a Female Apprentice and a Merchant Tailor in Ante-Bellum New England," *Labor History* 24, no. 2 (1983): 259–73; Rhyne, "Woman in Industry," 279–310; Edith Abbott, *Women in Industry* (New York: Appleton, 1910), 218–40; Feldman, *Fit for Men*, 107. Harriet Martineau, however, thought the condition of needlewomen must be "already ameliorated" by the sewing machine, as "their lot really could not be made worse," "The Needlewoman: Her Health," *Once a Week* 3, no. 24 (1860): 595. On French needlewomen, see Judith G. Coffin, *The Politics of Women's Work* (Princeton: Princeton University Press, 1996), 46–120.

96. Jesse E. Pope, *The Clothing Industry in New York* (New York: B. Franklin, 1970), 22–43; James J. Kenneally, *Women and American Trade Unions* (St. Albans, VT: Eden Press, 1978), 11; Elizabeth F. Baker, *Technology and Woman's Work* (New York: Columbia University Press, 1964), 20–26.

97. Abbott, *Women in Industry*, 221.

98. Horace Greeley et al., *The Great Industries of the United States* (Hartford: J. B. Burr Hyde, 1872), 592.

99. Despite his revolutionary effect on Western dress, George Brummel died penniless in France, deserted by all but—fittingly enough—his laundress. Two of many biographies of this flawed genius are Charles M. Franzero, *The Life and Times of Beau Brummell* (London: A. Redman, 1958); and Samuel Tenenbaum, *The Incredible Beau Brummell* (South Brunswick, NJ: A. S. Barnes, 1967). See also C. Willett and Phillis Emily Cunnington, *The History of Underclothes* (London: M. Joseph, 1951), 123–34, 183. On the effect of this cleanliness trend on the laundry industry, see Arwen Mohun, *Steam Laundries* (Baltimore: Johns Hopkins University Press, 1999), 29–39.

100. William Ewers and H. W. Baylor, *Sincere's History of the Sewing Machine* (Phoenix: Sincere Press, 1970), 91.

101. James Parton, *History of the Sewing Machine* (Boston: Howe Sewing Machine, 1868), 26; Ross Thomson, *The Path to Mechanized Shoe Production in the United States* (Chapel Hill: University of North Carolina Press, 1989), 73–117.

102. Harry Inwards, *Straw Hats* (London: Sir I. Pitman & Sons, 1922), 13–74, 117–23; C. T. Hinckley, "Making Artificial Flowers," *Godey's Lady's Book* 48 (1854): 295; W. H. Rideing, "Making Artificial Flowers," *Appleton's* 20:97; J. G. Austin, "Kings' Crowns and Fools' Caps," *Atlantic* 22 (1868): 428–39; Charlotte Rankin Aiken, *Millinery* (New York: Ronald Press, 1922), 49; John G. Dony, *A History of the Straw Hat Industry* (Luton, UK: Gibbs, Bamforth & Co., Leagrave Press, 1942), 114–15; Penny, *Employments of Women*, 292, 315, 337–45; and Harry B. Weiss and Grace M. Weiss, *The Early Hatters of New Jersey* (Trenton: New Jersey Agricultural Society, 1961), 35–46; and Edward M. Woolley, *A Century of Hats and the Hats of the Century* (Danbury, CT: Mallory Hat Company, 1923), 20.

103. Quoted in Harbeson, *American Needlework*, 150.

104. Albert Sidney Bolles, *Industrial History of the United States*, 3rd ed. (Norwich, CT: Henry Bill Publishing Co., 1878), 245; Ruth Brandon, *A Capitalist Romance: Singer and the Sewing Machine* (Philadelphia: Lippincott, 1977), 117–19; and Andrew B. Jack, "Channels of Distribution for an Innovation: The Sewing Machine Industry in America, 1860–1865," *Explorations in Entrepreneurial History*, o.s., 9, no. 3 (1957): 119.

105. Brian Jewell, *Veteran Sewing Machines: A Collector's Guide* (South Brunswick, NJ: A. S. Barnes, 1975), 16, 60; and Clark Thread Company, *A Thread Mill Illustrated* (Newark, NJ: Clark Thread Co., c. 1882). This product was still called "ONT" in 1956, nearly a century after its introduction; see *Priscilla Doilies to Crochet*, Coats & Clark's O.N.T. Book no. 324 (New York: Coats & Clark, 1956). Jewell is, of course, taking a somewhat narrow view of the term *mechanism*.

106. Grace Rogers Cooper, *The Sewing Machine: Its Invention and Development*, 2nd ed. (Washington, DC: Smithsonian Institution Press, 1976), vii.

107. The needle trades remain labor-intensive and low-wage, just as they were when Mary Harris Jones worked as a dressmaker in Chicago in the 1860s. Mary Harris Jones and Mary F. Parton, *The Autobiography of Mother Jones* (Chicago: C. H. Kerr for the Illinois Labor History Society, 1972), 13 and 115–16. For a modern view, see Jane Lou Collins, *Threads* (Chicago: University of Chicago Press, 2003).

108. See, for example, Singer Sewing Machine Company, *Singer Instructions for Art Embroidery and Lace Work* (1931; facsimile ed., Menlo Park, CA: Open Chain Pub., 1989).

109. Eliza C. Hall, *Aunt Jane of Kentucky* (New York: A. L. Burt, 1907), 58–59.

110. Jonathan Holstein, *American Pieced Quilts* (New York: Viking Press, 1973), 7–16.

111. Therle Hughes, *English Domestic Needlework, 1660–1860* (New York: Macmillan, 1961), 135–53; Susan B. Swan, "Quiltmaking within Women's Needlework Repertoire," in *In the Heart of Pennsylvania: Symposium Papers*, ed. J. Lasansky (Lewisburg, PA: Oral Traditions Project, 1986), 14.

112. For a more detailed discussion of textile mechanization, see Florence M. Montgomery, *Printed Textiles* (New York: Viking Press, 1970), and David Jeremy, *Transatlantic Industrial Revolution* (North Andover, MA: Merrimack Valley Textile Museum, 1981).

113. Bernard Rudofsky, *Are Clothes Modern?* (Chicago: P. Theobold, 1947), 137–50.

114. Harbeson, *American Needlework*, 38–39.

115. James A. B. Scherer, *Cotton as a World Power* (New York: Frederick A. Stokes Co., 1916), 44; and John Irwin and Katharine B. Brett, *Origins of Chintz* (London: H.M.S.O., 1970). On the adaptation of Asian-inspired designs in France, see Henri Clouzot and

Frances Morris, *Painted and Printed Fabrics* (New York: Metropolitan Museum of Art, 1927). Design influences flowed in the other direction as well; see Margaret Jourdain and Soame Jenyns, *Chinese Export Art in the Eighteenth Century* (London: Country Life, 1950), 61–64. For examples of home-made imitations of the palampore, see Carleton L. Safford and Robert C. Bishop, *America's Quilts and Coverlets* (New York: Dutton, 1972), 12–15, 40–85; Patsy Orlofsky and Myron Orlofsky, *Quilts in America* (New York: McGraw-Hill, 1974), 226–28; Rose Wilder Lane, *Woman's Day Book of American Needlework* (New York: Simon and Schuster, 1963), 40–49; and Mildred Davis, *The Art of Crewel Embroidery* (New York: Crown Publishers, 1962), as well as the same author's *Early American Embroidery Designs*.

116. Orlofsky and Orlofsky, *Quilts in America*, 94–96 and plate 89.

117. Diane L. Fagan Affleck, *Just New from the Mills* (North Andover, MA: Museum of American Textile History, 1987); Pettit, *America's Printed and Painted Fabrics, 1600–1900*, 38, 53–76, 172–215.

118. David J. Jeremy, *Transatlantic Industrial Revolution* (North Andover, MA: Merrimack Valley Textile Museum; Cambridge, MA: MIT Press, 1981).

119. The early manufacture of cotton thread in the United States is traditionally, but probably erroneously, associated with Samuel Slater. See, for example, George S. White, *Memoir of Samuel Slater* (Philadelphia: n.p., 1836); William R. Bagnall, *Samuel Slater and the Early Development of the Cotton Manufacture in the United States* (Middletown, CT: n.p., 1890); and the *New York Transcript* for November 1831, which credits Hannah Slater with the invention. See also Fennelly, *Textiles in New England, 1790–1840*, 22.

120. Hughes, *English Domestic Needlework*, 148.

121. Alice M. Earle, *Home Life in Colonial Days* (New York: Macmillan, 1898). Earle was, of course, wrong about the colonial dates for American patchwork quilts. For examples of the "revels" to which she and Hughes refer, see Safford and Bishop, *America's Quilts and Coverlets*, 96, 215, 217 and elsewhere in the color plates; and Orlofsky and Orlofsky, *Quilts in America*, 223–37, 297; as well as hundreds of other sources.

122. For examples, see Orlofsky and Orlofsky, *Quilts in America*, 237; and Rachel T. Pellman and Kenneth Pellman, *The World of Amish Quilts* (Intercourse, PA: Good Books, 1984).

123. Barbara Weiland, *Needlework Nostalgia: A Collection of Authentic Needlework Designs from the Butterick Archives* (New York: Butterick, 1975).

124. On the cutting of quilt patches, see Ruth E. Finley, *Old Patchwork Quilts and the Women Who Made Them* (Newton Centre, MA: Charles T. Branford, 1929), 57–60.

125. For examples of this type of construction, see Orlofsky and Orlofsky, *Quilts in America*, 107; and Lenice I. Bacon, *American Patchwork Quilts* (New York: Morrow, 1973), 139–43.

126. For examples of quilting templates, see Orlofsky and Orlofsky, *Quilts in America*, 153, and the discussion in the same source on pp. 171–85.

127. The character Patience Peabody, in the anonymous 1901 play "Aunt Jerusha's Quilting Party," in *Quilt Stories*, ed. C. Macheski (Lexington: University Press of Kentucky, 1994), 103, complains of a condition resembling carpal tunnel syndrome.

128. Charlotte M. Yonge, *Hopes and Fears* (Leipzig: B. Tauchnitz, 1861), 2:143.

129. Carrie A. Hall and Rose G. Kretsinger, *The Romance of the Patchwork Quilt in*

America (Caldwell, ID: Caxton Printers, 1935), 7. On the collection, see Bettina Havig, Carrie A. Hall, and Helen Foresman, *Carrie Hall Blocks: Over 800 Historical Patterns from the Collections of the Spencer Museum of Art, University of Kansas* (Paducah, KY: American Quilter's Society, 1999).

130. The expression "labor of obsession" is from Adele Aldridge, *Changes: A Book of Prints Inspired by the I Ching* (Riverside, CT: Mandala Press, 1972), introductory leaf.

131. Helen Campbell, *The American Girl's Home Book of Work and Play*, 2nd ed. (New York: Putnam, 1888), 234–47, 271.

132. Frances M. Trollope, *Domestic Manners of the Americans* (1839; London: Century; Toronto: Lester & Orpen Dennys Deneau, 1984), 42; see also Cowan, *More Work for Mother*, 26.

Chapter 4 · *The Hedonizing Marketplace*

1. Conference on Price Research, Committee on Textile Price Research, *Textile Markets* (New York: National Bureau of Economic Research, 1939), 67–87.

2. Examples in the collection of Colonial Williamsburg include accession nos. G1971-1283 and 1974-68.

3. Colonial Williamsburg has at least two such holders: accession no. G1971-1920 and a French example thought to date from the 1830s, G1971-3016.

4. Sylvia Groves, *The History of Needlework Tools and Accessories* (Feltham, UK: Hamlyn Publishing Group, 1966), 101.

5. Colonial Williamsburg's Abby Aldrich Rockefeller Folk Art Museum has a good collection of British, European, and American thimbles; G1971-3407-8 is only one of many examples.

6. "The Help Page," *Olde Time Needlework Patterns & Designs* (October/November 1978), 55. Historical examples of crochet hooks and knitting needles in standardized sizes can be found in the Smithsonian Museum of American History's collections (accession nos. T11182, T17861, H38995, T18422, and T18423-C) and in a group of artifacts that was unaccessioned at the time I examined it in the mid-1970s called "Drake 12-V-71."

7. *A Winter Gift for Ladies* (Philadelphia: B. G. Zieber, 1847), 15, 17. *Godey's Lady's Book* 58 (June 1859) had patterns for netting, *broderie anglaise*, and patchwork, and ibid., 80–81 (1870), designs for crochet, knitting, embroidery, net beadwork, "etching embroidery," braiding, and eyelet work. See ibid., p. 129, for a pattern for Cluny crochet, a heavy French style, and p. 187 for one for point Russe. On yarn weights for crochet and knitting, see Miss H. Burton, *The Lady's Book of Knitting and Crochet* (Boston: J. H. Symonds, 1875); Marie L. Kerzman, *Knitting* (Brooklyn: H. Bristow, 1884), 8. The needle size for rug knitting appears on p. 79 of Kerzman's book. The same author published a manual for crocheting the previous year, *Crochet Series Nos. 1–5* (Brooklyn: H. Bristow).

8. Barbara Walker, arguably America's best-known hand-knitting engineer, is the author of *A Treasury of Knitting Patterns* (1968), *The Craft of Lace Knitting* (1971), *Charted Knitting Designs* (1972), *Knitting from the Top* (1972), and *Sampler Knitting* (1973), all published by Scribner; and *A Fourth Treasury of Knitting Patterns* (Pittsville, WI: Schoolhouse Press, 2001), among other titles.

9. There is an early example of such a gauge in the collection of Colonial Williamsburg, accession, no. G1974-51.

10. On millinery establishments as needlework retailers, see *Needlecraft* 11, no. 2 (October 1919).

11. On six-cord, see *Cotton Goods Guide for Buyer and Seller* (New York: Dry Goods Chronicle Press, 1892), 163ff.; on the price of sewing thread, see *Silk* (New London, CT: Brainerd & Armstrong, 1909), 1–6.

12. Virginia Penny, *How Women Can Make Money* (1870; New York: Arno, 1971): 124–28.

13. *Lowell Directory* (1890), 966; *Lawrence Directory, Massachusetts* (1868–69), 244, 250; and ibid. (1900–1901), 745. All of these directories were published by Sampson & Murdock in Boston.

14. *Manchester, New Hampshire, Directory* (Boston, MA: Sampson & Murdock, 1881–1930); and *Pawtuxet Valley Directory, Rhode Island* (Boston, MA: Union Publishing, 1894). Peter Knights notes that city directories probably understate the number of these small businesses, as census data include more seamstresses than are listed in the directories. *Plain People of Boston, 1830–1860* (New York: Oxford University Press, 1971), 138.

15. This transition is documented in Gladys Thompson, *Patterns for Guernseys, Jerseys, and Arans*, 2nd ed. (New York: Dover, 1971).

16. On the continuing process of standardization, see *Standards and Guidelines for Crochet and Knitting* (Gastonia, NC: Craft Yarns Council of America, 2005), available at www.yarnstandards.com/s-and-g.pdf (accessed 28 July 2006).

17. *The Ladies' Self Instructor in Millinery and Mantua Making, Embroidery and Appliqué*, rev. ed. (Mendocino, CA: R. L. Shep, 1988; originally published in *Godey's Lady's Book* in 1853); and *The Self-Instructor in Silk Knitting, Crocheting, and Embroidery* (New York: Belding Bros., 1883). This latter was printed in an edition of 100,000 copies, suggesting that the manufacturer/publisher expected a ready market. Four more editions followed in quick succession.

18. References in Anthony Trollope, *The Kellys and the O'Kellys* (1845; Oxford: Oxford University Press, 1982), 169, 322, 372, 394, and 414, suggest that "worsted work" was not without its critics even at the height of its popularity.

19. Georgian B. Harbeson, *American Needlework* (New York: Bonanza Books, 1938), 107–12, 119–25.

20. Lilo Markrich and Heinz E. Kiewe, *Victorian Fancywork* (Chicago: Regnery, 1974).

21. M. E. Braddon, *The Doctor's Wife* (Leipzig: B. Tauchnitz, 1864), 2:133.

22. Margaret Oliphant, *He That Will Not When He May* (Leipzig: B. Tauchnitz, 1881), 1:38.

23. Lorine Pruette, *Women and Leisure* (New York: E. P. Dutton, 1924), 84.

24. Anna S. Richardson, "For the Girl Who Earns Her Own Living," *Woman's Home Companion* 32, no. 10 (1903): 40; "How I Earned Support for Others," *Woman's Home Companion* 32, no. 7 (1905): 38–39; "What Women Can Do," *Woman's Home Companion* (Aug. 1908): 5; "Does Needlework Pay?" *Needlecraft* 7, no. 12 (1916): 3; J.E.D., "A Bit of Experience," *Needlecraft* 10, no. 11 (1919), 3; and Margaret Barton Manning, "What Shall She Do?" *Needlecraft* 12, no. 2 (1920): 3.

25. Marianne Stradal, "Therese de Dillmont (1846–1890)," *Textilkunst* 3 (1979), available online at www.annatextiles.ch/biograph/dillmont/dillmo.htm (accessed 29 July 2006).

26. The non-English editions were *Encyclopédie des ouvrages de dames* (French), *Encyklopaedie der Weiblichen Handarbeiten* (German), *Enciclopedia de Labores de Señora* (Spanish), and *Enciclopedia dei Lavori Femminili* (Italian). WorldCat/OCLC currently lists twelve editions in English, of which the most recent is *The Complete Encyclopedia of Needlework*, anniversary ed. (Philadelphia: Running Press, 2002).

27. Thérèse de. Dillmont, *Polnyi Kurs Zhenskikh Rukodelii* (Moscow: EKSMO, 2004).

28. The English editions of these works in the DMC Library series published in Mulhouse (France before 1919 and after 1945, and Alsace in the interwar years) are *Czecho-Slovakian Embroideries* (undated); *Cross Stitch: New Designs*, 4th ser. (c. 1900); *Knotted Fringes* (undated); *Drawn Thread Work*, 1st and 2nd ser. (1900 and 1920); *Cross Stitch*, 7th ser. (c. 1900); *Teneriffe Lace Work* [and] *Crochet Work*, 1st ser. (1900); *Embroidery on Net*, 2nd ser. (c. 1900); *Colbert Embroideries* (1918); *Motifs for Embroideries* (1919); *Hardanger Embroideries* (1920); *Cross Stitch: New Designs*, 3rd ser. (1920); *Embroidery on Net* (1920); *Assisi Embroideries* (1920); *Embroiderer's Alphabet* (1920); *Hardanger Embroideries*, 2nd ser. (1921); *Jugoslavian Embroideries* (1950); *Bulgarian Embroideries* (1950); *Morocco Embroideries* (1955); *Tatting* (1956); *Cross Stitch*, 7th ser. (1961); *Openwork Embroidery*, 2nd ed. (1966); *Motifs for Embroideries* (1968); *Cross Stitch*, 5th ser. (1968); *Macrame* (1971); *Chinese Embroideries* (1977); and *Masterpieces of Irish Crochet Lace* (New York: Dover, 1986)

29. For examples of this new breed of lavishly illustrated and laid-out needlework publications, see Emma Farnes, *The Antique Design or Spider in Many Variations* (St. Louis, MO: E. C. Spuehler, [c. 1915]); Hugo W. Kirchmaier, *Book of Filet Crochet and Cross-Stitch*, Book, no. 6 (Toledo, OH: Cora Kirchmaier, 1919), or any other title in this series by the Kirchmaier design team, including *The Kirchmaier Book of Cross-stitch and Crochet in Color* (1914); Hugo W. Kirchmaier and Anna Wuerfel Brown, *The New Filet Crochet Book* (Toledo, OH: C. Kirchmaier, 1912); and numbers 4 and 5 of this series, both published in 1915. See also Mrs. F. W. Kettelle, *The Priscilla Filet Crochet Book, no. 2.* (Boston: Priscilla Publishing Co., 1914).

30. My list of needlework titles draws on "Anhang zu: Historische Frauenzeitschriften," compiled by Christa Bittermann-Wille and Helga Hofmann-Weinberger for the Beitrag für das Projekt *KolloquiA - Forschungs- und Lehrmaterialien zur Frauenrelevanten und Feministischen Dokumentations- und Informationsarbeit in Österreich*, Projektleitung: Helga Klösch-Melliwa, gefördert vom Jubiläumsfonds der Österreichischen Nationalbank (Nr. 6816), vom BM für Wissenschaft und Verkehr und vom BM für Arbeit, Gesundheit und Soziales, available at www.onb.ac.at/ariadne/pubhistzbib.htm (accessed 2 Aug. 2006), among other sources.

31. S. Giedion, *Mechanization Takes Command* (New York: Oxford University Press, 1948), 258–510.

32. Hazel H. Adler, "Planning the Furnishing of Your Home," *Needlecraft* 18, no. 1 (1926): 6–7; Grace Hall Stratton, "We Create Modern Needlepoint Loveliness with the Old-time Tent-Stitch," *Needlecraft* 19, no. 5 (1928): 12; and International Heater Company, *International Onepipe Heater Makes Your Entire Home Warm, Cozy, and Comfortable* (Utica: The Company, 1920).

33. U.S. Bureau of the Census, *Historical Statistics of the United States: Colonial Times to 1970* (Washington, DC: Government Printing Office, 1975), 1:164–65.

34. *Needlecraft* 10, no. 4 (1918): 2.

35. T. M. James, *Longmans' Complete Course of Needlework, Knitting, and Cutting Out* (London: Longmans, Green, 1901). Longmans published textbooks on many topics, including geography and "Sea Dayak" arithmetic. See also Idabelle McGlauflin, *Handicraft for Girls* (Peoria, IL: Manual Arts Press, 1915); Margaret Swanson, *Needlecraft in the School* (London: Longmans, Green, 1916); the same author's *Sewing Handicraft for Girls: A Graded Course for City and Rural Schools*, rev. ed. (Peoria, IL: Manual Arts Press, 1918); Ellen P. Claydon and C. A. Claydon, *Knitting without "Specimens"* (New York: E. P. Dutton, 1915); Carrie Syphax Watson, *Notes on First Grade Sewing* (Rochester, NY: Mechanics Institute, 1903); and Mrs. L. Floyer with C. S. Watson, *Educational Sewing Squares Printed with Designs for Embroidery, Guides for Basting and Darning . . . Plain Knitting and Mending, in Six Standards* (London: School Board, 1876); and *Home Sewing: Embroideries and Findings* (Scranton, PA: Woman's Institute of Domestic Arts and Sciences, 1923).

36. Mary Whitman, "How I Earn Money at Home" [advertisement for Auto Knitter Hosiery Co. of Buffalo NY], *American Needlewoman* 35, no. 6 (1925): 2; and Ida Riley Duncan, *Knit to Fit* (New York: Liveright, 1963).

37. Quoted in Catherine Fennelly, *Textiles in New England, 1790–1840* (Sturbridge, MA: Old Sturbridge Village, 1961), 3. See also a fictional reference to this kind of renunciation of embroidery in Anya Seton, *Katherine* (Chicago: Chicago Review Press, 2004), 126.

38. See, for examples, *Directions for Knitting the Army Mitten* (Washington, DC: U.S. Army, 1861); Robert P. Weston, Herman Darewski, and Jack Norworth, *Mother's Sitting Knitting Little Mittens for the Navy* [sheet music] (London: Francis, Day, & Hunter, 1915); "European Relief," *Needlecraft* 10 (Feb. 1919): 3; Edna Goodheart, "Comforts for Our Soldiers," *Needlecraft* 9, no. 6 (1917): 6; "War Work for Willing Hands," *Star Needlework Journal* 3, no. 1 (1918): 4; and American National Red Cross, *Red Cross Service Record: July 1, 1939–June 30, 1946* (Washington, DC: American National Red Cross, 1946), 9. There also an historical account of knitting in this context in Anne L. Macdonald, *No Idle Hands* (New York: Ballantine Books, 1988), 26–43, 97–133, 199–238, and 289–320. Mollie Panter-Downes observed in 1947 that "men must fight and women must sew" in the context of how satisfying it was to sit in a group of women doing needlework for the war effort, in *One Fine Day* (1947; London: Virago, 1985), 112; see also John Galsworthy, *End of the Chapter* (New York: Scribner's, 1935), 373. For a Freudian interpretation of this kind of labor, see M. J. Farrell, *Loving without Tears* (1951; London: Virago, 1988), 27.

39. On Dorcas work, see, for example, Florence Marryat, *Petronel* (Leipzig: B. Tauchnitz, 1870), 1:269–70. Other references, including fictional portrayals of charity needlework in this context include Isabel Allende, *The House of the Spirits* (New York: Knopf, 1999), 238; Barbara J. Morris, *Victorian Embroidery* (London: H. Jenkins, 1962), 32–49; Laura Talbot, *The Gentlewomen* (1952; London: Virago, 1985), 53; E. H. Young, *The Curate's Wife* (1934; New York and London: Penguin and Virago, 1985), 93; Margaret Oliphant and Thomas B. Aldrich, *The Second Son* (Boston: Houghton Mifflin, 1888), 63; and M. E. Braddon, *The Doctor's Wife* (Leipzig: B. Tauchnitz, 1864), 1:97, 148–49. On needlework for charity fundraising, see also Samuel Orchart Beeton, *The Lady's Bazaar and Fancy Fair Book*

(London: Ward Lock, [c. 1880]); and Beverly Gordon, *Bazaars and Fair Ladies: the History of the American Fundraising Fair* (Knoxville: University of Tennessee Press, 1998).

40. Cornelia Mee and Austin Mee, *The Work-Table Magazine of Church and Decorative Needlework* (London: D. Bogue, 1847); *Emblems and Church Laces*, Star Book, no. 50 (New York: American Thread Company, 1947); Blanche Saward's section on church needlework in Sophia Frances A. Caulfeild and Blanche C. Saward, *The Dictionary of Needlework*, 2nd ed. (London: Gill, 1887); Beryl Dean, *Ideas for Church Embroidery* (Newton Center, MA: Branford, 1968); and Lillian S. Freehof and Bucky King, *Embroideries and Fabrics for Synagogue and Home* (New York: Hearthside Press, 1966). There is a fictional portrayal of an altar guild sewing circle in Dorothy CanfieldFischer, *The Home-Maker* (New York: Harcourt Brace, 1924), 56. For religious embroideries used in the home, see Catherine Schwoeffermann, Peter Klosky, and Merrill Oliver, *Goddesses and Their Offspring: Nineteenth- and Twentieth-Century Eastern European Embroideries* (Binghamton, NY: Roberson Center for the Arts and Sciences, 1986), 14, 36–41.

41. *Florence Home Needlework* (Florence, MA: Nonotuck Silk Co., 1894); and Mrs. F. W. Kettelle, *The Priscilla Filet Crochet Book, no. 2* (Boston: Priscilla Publishing, 1914), 38. Kettelle's work has been reprinted as *Filet Crochet: Projects and Designs* (New York: Dover, 1979).

42. For early-twentieth-century examples of the popularity of hardanger in the United States, see Margaret Barton Manning, *Handbook of Needlecraft*, no. 2 (Augusta, ME: Needlecraft Publishing, [c. 1920]), 54–56; Mary E. Fitch, "A Hardanger Centerpiece," *Needlecraft* 24, no. 10 (1923): 22; Sophie T. La Croix, *Old and New Designs in Hardanger Embroidery* (St. Louis: St. Louis Fancy Work Co., [c. 1900]); "Collar and Cuff Sets in Hardanger Embroidery," *Home Needlework Magazine* 6, no. 4 (1904): 331–33; E. D. Moerke, "Designs in Hardanger Embroidery," *Needlecraft* 7, no. 5 (1916): 13; Anna M. Porter, *Norwegian Drawn Work (Hardanger)* (Belmar, NJ: The Author, 1904); Sara Hadley, *The Complete Hardanger Book* (New York: The Author, 1904); *Healy Hardanger Samplers* (Detroit: Healy, [c. 1915]; and Nellie Clarke Brown, "Hardanger Centerpiece in Green and White," *Needlecraft* 16, no. 12 (1934): 7.

43. Lilian Barton Wilson, "Danish Hedebo in Coarse Linen Thread," *Home Needlework Magazine* 27, no. 8 (1915): 6–8; "Hedebo and Crochet in Good Accord," *Plain and Fancy Needlework* 2, no. 12 (1917): 12; *The Priscilla Manual* (Boston: Priscilla, 1905); and "Hedebo Embroidery," *Woman's Weekly Home Arts and Entertainment Annual Supplement* (1922), 33.

44. Flora Klickmann, *The Cult of the Needle* (New York: F. A. Stokes, 1915).

45. Scientific racism and ethnic prejudices flourished in this period; see, for example, Frederick L. Hoffman, *Race Traits and Tendencies of the American Negro* (New York: American Economic Association by Macmillan Co., 1896). The work is a classic of arrogant delusion, predicting the downfall and probable extinction of all "races" except the author's own, the Aryan.

46. See, for example, "Italy," in Richard Mullen and James Munson, *The Penguin Companion to Trollope* (London: Penguin, 1996), 240–41.

47. John Q. Reed and Eliza M. Lavin, *Needle and Brush* (New York: Butterick Publishing Co., 1889), 13–18, 19–25; Louisa A. Tebbs, *The New Lace Embroidery (Punto Tagliato)*, 2nd ed. (London: Chapman & Hall, 1905); Vera Best, "Italian Lace," *Needlecraft* 11, no. 1

(1919): 22; "Beautiful Luncheon Linens in Italian Embroidery," *Modern Priscilla* 36, no. 12 (1923): 5; and Lilian Barton Wilson, "Handsome Tassels for Italian Embroideries," ibid., pp. 6–7.

48. See, for examples, "Specially Priced 69 Cents: Apron 2534," *American Needlewoman* 35, no. 6 (1925): 21; "Butterfly-Design Hemstitched Scarf," *Good Stories* 43, no. 10 (1927): 31; "Cross-Stitch Basket Design: Tan Linen Fringed Scarf," *Needlecraft Gift Book* 20 (1929): 10; and "Rose Design Pillow Slips on 42-Inch Tubing Hemstitched for Crocheting," ibid., p. 20.

49. On these searches, see Florence Yoder Wilson, "We Find Embroideries in Yugoslavia," *Needlecraft* 24, no. 2 (1932): 4, and her "Costume Hunting in Dalmatia," ibid., p. 16.

50. Carmela Testa, *Variety* (Boston, MA: Carmela Testa & Co., c. 1917–1920s), and reprints available at Iva Rose Vintage Reproductions, www.ivarose.com/carmela_testa?b=1 (accessed 9 Aug. 2006). See also Anna Ferrari, "Reproduction of a Quaint Buffet Scarf in Italian Filet," *Needlecraft* 13, no. 12 (1922): 8. An example of Rita Garcia's work is "Colado Philippine Punchwork," *Needlecraft* 9, no. 6 (1918): 7. In the Philippines, embroidery and lace were industries, not hobbies; see Philippines Bureau of Education, *Lace Making and Embroidery* (Manila: Bureau of Print, 1911).

51. See, for examples, Vera Best, "Italian Lace," *Needlecraft* 11, no. 1 (1919): 22; Lillian Baynes Griffin, "Porto Rican Lace Work," *Home Needlework Magazine* 4, no. 4 (1902): 318–22; Margaret Barton Manning, "A Handsome Square in Mexican Work," *Handbook of Needlecraft*, no. 2 (Augusta, ME: Needlecraft Publishing, [c. 1920]), 51; and "Mexican Drawnwork," *Needlecraft* 8, no. 3 (1916): 12.

52. A few examples are Anne Champe Orr, *A Book of Baby Caps and Nursery Designs in Crochet Work* (Nashville, TN: Anne Orr Studio, 1915); *Center Pieces and Lunch Sets in Crochet Work* (Nashville: Anne Orr Studio, 1915); *Pictorial Review Book of Crocheting and Knitting*, vol. 5 (New York: Pictorial Review, 1920); *Yokes, Sweaters, and Lingerie* (Nashville: Anne Orr Design Co., 1921); *J. & P. Coats Centerpieces and Edgings*, Book 16 (Pawtucket, RI: J. & P. Coats, 1923); *Star Book of Crochet Designs*, no. 1 (New York: American Thread Co., 1935); *Weave-it Afghans* (Middleboro, MA: Hero Mfg. Co., 1937); *Creative Crochet with Carpet Warp and Candlewick Cotton* (New York: Hooker & Sanders, 1940); and *Tatting* (Nashville: Anne Orr Studio, 1942).

53. Anne Champe Orr, *Crochet Designs of Anne Orr* (New York: published for the Center for the History of American Needlework by Dover, 1978), and in the same Dover series, *Anne Orr's Charted Designs* (1978); *Favorite Charted Designs* (1983); and *Anne Orr's Filet Crochet Designs* (1986). See also Jean DuBois, "Anne Orr: She Captured Beauty," *Quilters' Newsletter* 91 (1977): 12–14, 27; and advertisements in *Needlecraft* 11, no. 3 (1919): 17, and 12, no. 4 (1920): 3.

54. Elizabeth Boyle, *The Irish Flowerers* (Holywood and Belfast: Cultra Manor, Ulster Folk Museum, and Queen's University Institute of Irish Studies, 1971); for adaptations, see "Irish Tatting," *Woman's Weekly Home Arts and Entertainment Annual Supplement* (1922), 19; Anna Hoskins, "Coat Collar in Tatting," *Needlecraft* 10, no. 8 (1919): 22; and Sara Hadley, "Irish-Point," *Lace Maker* 2, no. 6 (1904). This journal also published patterns for Irish crochet and for Carrickmacross, a hybrid technique of appliqué embroidery on machine-made net lace.

55. See, for example, Mary E. M. Fitch, *Filet Crochet with Instructions*, 2nd ser. (Brookline, MA: The Author, 1915); Anna Valeire, *Finished Yokes of Beauty* (St. Louis: Spuehler, 1920); Belle Robinson, *The Priscilla Filet Crochet Book* (Boston, MA: Priscilla Publishing Co., 1915); and Elsa Barsaloux, *Richardson's Cross-Stitch Book and Filet Crochet* (Chicago: Richardson Silk Co., 1916).

56. Anne Orr, *J. & P. Coats Crochet Book, No. 2* (Pawtucket, RI: J. & P. Coats, 1917); Anna W. Brown, "Neck Accessories in Irish Crochet," *Home Needlework Magazine* 12, no. 2 (1910): 105; Frances D. Johnson, "New Design for an Irish Crochet Collar," *Home Needlework Magazine* 13, no. 3 (1911): 175; Mary Card, "Irish Crochet," *Needlecraft* 12, no. 1 (1920): 10; the same author's "A First Lesson in Irish Crochet," *Needlecraft* 11, no. 12 (1920): 8–10; A.G.F., "A Bit of Inspiration" [on Mary Card's interpretation of Irish crochet], *Needlecraft* 9, no. 7 (1918): 3. Mary Card was an Australian-born teacher who may have had Irish antecedents. Margaret Barton Manning, "Introducing Mary Card," *Needlecraft* 9, no. 6 (1918): 3. See also Card's "Counterpane of Knitting and Linen," *Needlecraft* 11, no. 10 (1920): 5, and an advertisement for her patterns in *Modern Priscilla* 36, no. 8 (1922): 70.

57. "Mountmellick Work," *Home Needlework Magazine* 3, no. 2 (1901): 131; and *Weldon's Practical Mountmellick Embroidery*, 4th ser. (London: Weldon's, [c. 1900]). *Home Needlework* included Mountmellick patterns in almost every issue at this period.

58. On the Armenian genocide and international aid efforts, see Christopher J. Walker, *Armenia: The Survival of a Nation* (New York: St. Martin's Press, 1980); Merrill D. Peterson, *"Starving Armenians"* (Charlottesville: University of Virginia Press, 2004); and J. M. Winter, *America and the Armenian Genocide of 1915* (Cambridge: Cambridge University Press, 2003). On Armenian needlework, see Alice O. Kasparian, *Armenian Needlelace and Embroidery* (McLean, VA: EPM Publications, 1983), and Serik Davt`yan, *Haykakan Zhanyak* (Erevan, Armenia: HSSR GA Hratarakch`ut`yun, 1966).

59. Barbour Flax Spinning, "Armenian Edge—Easily Worked," in *Barbour's Linen Thread for Art Needlework and Crocheting* (Paterson, NJ: Barbour Flax Spinning, n.d.); "One Makes This Lace with a Sewing Needle: Armenian Needlepoint Stitches," *Star Needlework Journal* 10, no. 1 (1925): 13; Marie Haase, "Armenian Lace and Little Embroidery," *Star Needlework Journal* 10, no. 3 (1925): 6; Nouvart Tashjian, "Armenian Needle-Point Lace," *Plain and Fancy Needlework* 2, no. 8 (1917): 5, and ibid., 2, no. 12 (1917): 7; Nouvart Tashjian, "Armenian Lace Flower Pendants," *Modern Priscilla* 36, no. 8 (1922): 9; Nouvart Tashjian, Jules Kliot, and Kaethe Kliot, *Armenian Lace* (1923; Berkeley, CA: Lacis Publications, 1982); and Marie Haase, "A Lovely Centerpiece Simulating the Real Armenian Lace" [in crochet], in *Selected Old Time Needlework Patterns*, ed. E. Kutlowski (Danvers, MA: Tower Press, n.d.). See also the undated "Dress Trimmings in Armenian Lace," in Arlene Z. Wiczyk, *A Treasury of Needlework Projects from "Godey's Lady's Book"* (New York: Arco, 1972), 103.

60. Martha Stearns, *Homespun and Blue* (New York: Scribner, 1963), 65–67 and 83. Battenberg was a revival of the Italian Renaissance art of making lace from tape, reintroduced about 1885 with machine- rather than hand-woven tape as the base. Virginia C. Bath, "Tape Laces," in *Lace* (New York: Penguin, 1979), 292–96; *The Priscilla Manual* (Boston: Priscilla, 1905); Christine Perry, "Enter Battenberg Laces," *Needlecraft* 23, no. 11 (1932): 5; "Battenberg," *Woman's Home Companion* 32, no. 10 (1905); and Jennie T. Wandle, "Royal Battenberg Lace," *Home Needlework Magazine* 2, no. 2 (1900): 154–58.

61. Gale Literary Database's *Contemporary Authors* (accessed 12 Aug. 2006) mentions Stearns's Republican affiliation, her 1905 marriage to a future U.S. Congressman, and her memberships in the Colonial Dames of America, the New Hampshire League of Arts and Crafts, and the Boston Society of Arts and Crafts. She was born in Amherst, Massachusetts, the daughter of college professor John Franklin Genung and Florence Mable Sprague Genung. Stearns was also an embroidery designer; see, for example, her "Christmas Bags of Linen," *Modern Priscilla* 24, no. 10 (1910): 13.

62. Candace Wheeler, *The Development of Embroidery in America* (New York: Harper and Brothers, 1921), 102–20; Paul Jameson Woodward, *Catalogue of Early American Handicraft* (Brooklyn: Museum Press, 1924); Victoria and Albert Museum, Department of Textiles, *Catalogue of Samplers*, 2nd and 3rd eds. (London: H. M. Stationery Office, 1915 and 1922); Preston Remington, *English Domestic Needlework of the XVI, XVII, and XVIII Centuries* (New York: Metropolitan Museum of Art, 1945); Margaret E. White, *Quilts and Counterpanes in the Newark Museum* (Newark: Newark Museum, 1948).

63. Examples include Margaret L. Brooke and Winifred M. A. Brooke, *Lace in the Making with Bobbins and Needle* (London: Routledge, 1923); Gertrude Whiting, *Tools and Toys of Stitchery* (New York: Columbia University Press, 1928); Thomas Woodhouse, *The Handicraft Art of Weaving* (London: H. Frowde, 1921); Edward F. Worst, *Foot-Power Loom Weaving* (Milwaukee, WI: Bruce Publishing, 1918); Margaret Maidment, *A Manual of Hand-Made Bobbin Lace Work* (London and New York: Pitman and Pesel, 1931); *Practical Canvas Embroidery* (London: B. T. Batsford, 1929); Charlotte Kellogg, *Bobbins of Belgium* (New York: Funk & Wagnalls, 1920); Louisa A. Tebbs and Rosa Tebbs, *The Art of Bobbin Lace*, 3rd ed. (London: Chapman & Hall, 1911); C. Geoffrey Holme et al., *A Book of Old Embroidery* (London: "The Studio" Ltd., 1921); R. E. Head, *The Lace and Embroidery Collector* (London: H. Jenkins, 1922); Ernest Lefebure, *Embroidery and Lace* (London: H. Grevel, 1888); Emily Leigh Lowes, *Chats on Old Lace and Needlework* (London: T. F. Unwin, 1908); Emily Jackson and Ernesto Jesurum, *A History of Hand-Made Lace* (London and New York: L. Upcott Gill and C. Scribner's Sons, 1900); Lillian E. Simpson and M. Weir, *The Weaver's Craft*, 3rd ed. (Peoria, IL: Manual Arts Press, 1939); Elizabeth Mincoff and Margaret S. Marriage, *Pillow Lace* (New York: Dutton, 1907); and William W. Kent, *The Hooked Rug*, 2nd ed. (New York: Tudor Publishing Co., 1937).

64. Needlework examples of this British effort to create a "useable past" include Marianne M. C. Alford and Elizabeth G. Holahan, *Needlework as Art* (London: S. Low, Marston, Searle, and Rivington, 1886); A. G. I. Christie, *English Medieval Embroidery* (Oxford: Clarendon Press, 1938); Albert F. Kendrick, *English Embroidery* (London and New York: B. T. Batsford and C. Scribner's Sons, 1904); A. F. Kendrick, *English Needlework* (London: A. & C. Black, 1933); Louisa F. Pesel, *English Embroidery* (London: B. T. Batsford, 1931); and Grace Christie, *Samplers and Stitches* (London: B. T. Batsford, 1921).

65. For example, George Oprescu, *Peasant Art in Romania* (London: Studio Ltd., 1929); Eric Kolbenheyer, *Motive der Hausindustriellen Stickerei in der Bukowina* (Vienna: K. K. Hof- und Staatsdrukerei, 1912); Halfdan Arneberg, *Norwegian Peasant Art* (Oslo: Fabritius and Fael, 1949); and *Hungarian Peasant Embroidery* (London: B. T. Batsford, 1961). On the importance of traditional textiles to ethnic identification, see Joanna Dankowska, *Textiles and National Identity among Ukrulnians in Poland* (Pittsburgh: Center for

Russian & East European Studies, University Center for International Studies, University of Pittsburgh, 1996).

66. *The Art Journal Illustrated Catalogue* (London: G. Virtue, 1851); textiles appear on pp. 39, 42, 44, 53, 55, 63, and 69.

67. Eileen Boris, *Art and Labor* (Philadelphia: Temple University Press, 1986), 99–138; and Marianne Carlano and Nicola J. Shilliam, *Early Modern Textiles* (Boston: Museum of Fine Arts, 1993).

68. Joan Campbell, *The German Werkbund* (Princeton: Princeton University Press, 1978), 51–52, 157, 200, and 225.

69. Harbeson, *American Needlework,* 159–65; Wheeler, *Development of Embroidery in America,* 102–20; and Toledo Museum of Art, *The Art of Louis Comfort Tiffany* (Toledo: The Museum, 1978), 41–43, 76.

70. *Royal Society Tatting and Crochet Lessons,* Royal Society Crochet Books, no. 5 (New York: Royal Society Press, 1915); *Royal Society Crochet Lessons,* no. 6 (New York: H. E. Verran, 1915); *Royal Society Crochet and Knitting,* no. 15 (New York: H. E. Verran, 1920); *Royal Society Crochet Book,* no. 18 (New York: H. E. Verran, 1921); and many other titles along similar lines. The style of these patterns was as "Victorian" in appearance as those of *Godey's* in the previous century.

71. On the Colonial Revival, see Laurel T. Ulrich, *The Age of Homespun* (New York: Knopf, 2001), 11–40; and Beverly Gordon, "Spinning Wheels, Samplers, and the Modern Priscilla: The Images and Paradoxes of Colonial Revival Needlework," *Winterthur Portfolio* 33, nos. 2/3 (1998): 163–94.

72. Wheeler, *Development of Embroidery in America,* 113; Harbeson, *American Needlework,* 152–54; and Stearns, *Homespun and Blue,* 66–86.

73. The footprint of the original house, which burned in 1779, is still visible at the site. The association objected not only to the small size but also to the position of the original house, as offering no view of the water; this proved fortunate as it preserved the original site for later archaeology. See National Park Service, "Memorial House and Museum," www.nps.gov/gewa/Page13Bnewhouse.html (accessed 15 Aug. 2006); and J. Paul Hudson, *George Washington Birthplace National Monument, Virginia* (Washington, DC: National Park Service, 1956).

74. *Index of American Design* (Chicago: Federal Art Project, Works Progress Administration, 1936). For idealized views of needlework and other home arts in Colonial American life, see Clarence P. Hornung, *Treasury of American Design,* vol. 2 (New York: H. N. Abrams, 1972).

75. A few of hundreds of possible examples are Alice M. Earle, *Home Life in Colonial Days* (New York: Macmillan, 1898); Harold D. Eberlein, Abbot McClure, and Mabel F. Bainbridge, *The Practical Book of Early American Arts and Crafts* (Philadelphia: Lippincott, 1916); Jane T. August Robinson, "The Kitchen of the Colonial Home," *House and Garden* 46 (1924): 78–79; Mary E. Gould, *The Early American House* (New York: M. McBride, 1949); Burl N. Osburn and Bernice B. Osburn, *Measured Drawings of Early American Furniture* (Milwaukee: Bruce Publishing, 1926); Rhea M. Knittle, *Early American Glass* (New York: Century, 1927); and Frances Little, *Early American Textiles* (New York: Century, 1931).

76. Marianna M. Hornor, "Whitman Sampler Traveling Exhibition," *Needle Art / Embroiders' Guild of America (N.Y.)* 2, no. 3 (1971): 5–14.

77. The first books published on quilting in the United States were Carrie A. Hall and Rose G. Kretsinger, *The Romance of the Patchwork Quilt in America* (Caldwell, ID: Caxton Printers, 1935); Ruby S. McKim, *One Hundred and One Patchwork Patterns* (Independence, MO: McKim Studios, 1931); and Florence Peto, *American Quilts and Coverlets* (New York: Chanticleer Press, 1949). Susan Currell attributes some of this crafts-revival activity to the increase of leisure provided by Depression unemployment; see her *March of Spare Time* (Philadelphia: University of Pennsylvania Press, 2005), 70–74.

78. Ulrich, *Age of Homespun.*

79. National Restaurant Association Research Department, *Meal Consumption Behavior* (Washington, DC: National Restaurant Association, 2002).

80. Jerry Adler et al., "Takeout Nation," *Newsweek* 143, no. 6 (2004): 52.

81. Two examples from the seventeenth and eighteenth centuries, respectively, are Kenelm Digby, *The Closet of the Eminently Learned Sir Kenelme Digby kt. Opened,* 3rd ed. (London: H.C. for H. Brome, 1677); and *The Lady's Companion,* 6th ed. (London: J. Hodges, 1753).

82. For examples, see *The Good Hous-Wives Treasurie* (London: Edward Allde, 1588); and Joseph D. V. Beeton, *Beeton's Every-Day Cookery and Housekeeping Book* (London: Ward Lock, 1872).

83. American and British examples include A. Hausner, *The Manufacture of Preserved Foods and Sweetmeats* (N.p.: Scott Greenwood, 1912); William T. Brannt, *A Practical Treatise on the Manufacture of Vinegar,* 2nd ed. (Philadelphia and London: H. C. Baird and Sampson Lou Marston Co., 1900); C. A. Shinkle, *American Commercial Methods of Manufacturing Pickles, Preserves, Canned Goods, Etc.* (Baltimore: Trade Co., 1902); Joseph Coppinger, *The American Practical Brewer and Tanner* (New York: Van Winkle and Wiley, 1815); John Gardner, *The Brewer, Distiller, and Wine Manufacturer* (Philadelphia: P. Blakiston, 1883); and James Death, *The Defects of Beer: Their Causes and Remedies* (London: The Author, 1889).

84. See, for example, Mary Eales, *The Compleat Confectioner* (London: J. Brindley and R. Mountagu [sic], 1733); and Hannah Glasse, *The Compleat Confectioner* (London: Mrs Ashburner, Deard's I. Pottinger and J. Williams, 1760). The fact that these works have the same title suggests that Glasse may have hoped to draw on the reputation of an earlier successful work.

85. Henry Weatherley, *A Treatise on the Art of Boiling Sugar, Crystallizing, Lozenge-Making, Comfits, Gum Goods, and other Processes for Confectionery* (Philadelphia: H. C. Baird, 1865); Henry Weatherley and John Fuller, *Weatherley's Confectioner and Practical Guide,* 7th ed. (Rochdale, UK: E. Wrigley & Sons, 1890); Jacob Friedman, *Friedman's Common-Sense Candy Teacher* (Chicago: J. N. Bell, 1906); and W. O. Rigby and Fred Rigby, *Rigby's Reliable Candy Teacher with Complete and Modern Soda, Ice Cream, and Sherbet Sections,* 14th ed. (Topeka, KS: Rigby Publishing, 1923).

86. *Fifty Tested Recipes for Candy and Cake Icings* (N.p.: Home Candymaker's Thermometer, 1920). Traditional home preserving methods have included two tests for the jelly point other than by using a candy thermometer: sheeting of the cooked syrup from a spoon

and cooling a drop of it on a chilled plate. Both methods are described in Fannie M. Farmer, Marion Cunningham, and Jeri Laber, *The Fannie Farmer Cookbook*, 12th ed. (New York: Knopf, distributed by Random House, 1979), 700.

87. *Hood's Book of Home Made Candies* (Lowell, MA: C. I. Hood & Co., 1888); *Confectionery Is Always Most Acceptable When Artistic in Shape* (Boston: W. M. Baker, 1890); *Fine Candies Are Easy to Make with Karo Syrup* (New York: Corn Products Refining Co., 1900); *How to Make Xmas Candies and Dainties* (Johnstown, NY: Knox Gelatine, 1900); *Pennant Recipes for Tempting Foods* (Columbus, IN: Union Starch & Refining Co., 1900); *Franklin Sugar Candy Book* (Philadelphia: Franklin Sugar Refining Co., 1900); *Delicious Desserts and Candies*, 3rd ed. (Chicago: Price Flavoring Extract Co., 1923); *Perfect Chocolates of Your Own Making* (Dorchester, MA: Walter Baker & Co., Educational Dept., 1928); and *Candies Made from the Pet Milk Experimental Kitchens* (St. Louis, MO: Pet Milk Co., 1940) For an example of juvenile literature on this subject, see Elizabeth Du Bois Bache and Louise Franklin Bache, *When Mother Lets Us Make Candy* (New York: Moffat, Yard & Co., 1921).

88. *Sweets* (Lynn, MA: Lydia E. Pinkham Medicine Co., 1910); *Candies* (Milwaukee: Milwaukee Gas Light Co., 1933); J. E. Hissong, *The Marvel Candy Instructor* (Coshocton, OH: Marvel, 1909); and Rose Brown, *Sweets for Bazaars, Children's Holidays, &c.* (London: Simpkin, Marshall, 1905).

89. W. V. Cruess and Agnes O'Neill, *The Home Preparation of Fruit Candy* (Berkeley: University of California, 1927 and 1938); *Merry Christmas Candy* (New Brunswick, NJ: Extension Service, College of Agriculture, Rutgers University, 1951); Edwin J. A. Anderson, *Honey Candies, no. 186* (University Park, PA: Pennsylvania State University, College of Agriculture, 1958); and Evangeline J. Smith, *Candy* (Laramie: University of Wyoming, Agricultural Extension Service, 1963).

90. Therle Hughes, *Sweetmeat and Jelly Glasses: Antique Collectors Pocket Guides* (Guildford, Surrey: Lutterworth Press, 1982). Hughes, as we have seen, also writes about needlework for collectors; see chapter 3, note 30.

91. M. A. Carême, *Le Pâtissier Pittoresque* (Paris: F. Didot, 1815). This work, which emphasizes the construction of cakes that resembled buildings (including ruins), was evidently a success, going into a fourth edition by 1854 as *Le Pâtissier Pittoresque: Contenant cent Vingt-Cinq Planches Gravées au Trait, dont Cent Dix Représentent une Variété de Modèles de Pavillons, Rotondes, Temples, Ruines, Tours, Belvédères, Forts, Cascades, Fontaines, Maisons de Plaisance, Chaumières, Moulins et Ermitages: Précédé d'un Traité des Cinq Ordres d'Architecture, selon Vignole: Auquel on a Joint des Détails des Ordres Cariatide, Poestum, Égyptien, Chinois et Gothique, Tirés du Parallèle des Monuments Antiques et Modernes* (Paris: Dépôt de Librairie). The links of cake decorating with architecture and historical archaeology deserve at least a doctoral dissertation.

92. See, for examples, *Catalogue of Materials for Decorating Cakes, Pastry, Confectionery, Easter Eggs, etc* (Reading, PA: Stichler & Co., 1930); *Illustrated Catalogue of Bride Cakes: Special Designs for Naval and Military Weddings* (Plymouth, Eng.: H. Matthews & Sons, 1900); Herman Hueg, *Ornamental Confectionery and Practical Assistant to the Art of Baking* (1893); and John F. Schaer, *Confectioner's Guide* (Baltimore: The Author, 1856).

93. *Decorator Fun for Everyone* (Minneapolis: Pillsbury, 1960); Harriet Chelmo, *Let's Decorate a Cake* (Minneapolis: Pillsbury, 1957); *Betty Crocker's Frosting Secrets: Fun with*

Frostings (Minneapolis: General Mills, 1958); Orma N. Farnham, *Cake Decorating for Home-makers* (New York: William-Frederick Press, 1953); and Richard V. Snyder, *Decorating Cakes for Fun and Profit*, 7th ed. (New York: Exposition Press, 1960).

94. For example, the Rhode Island summer kitchen of my great-grandmother Ethel Richmond Niles, a late-nineteenth-century renovation to an older house, was located in the one-story "ell" addition to her two-story frame farmhouse, with multiple doors and windows to facilitate cross-ventilation. A hand-pump provided water. This part of the house was not heated in winter. My late mother-in-law, Alta Opal Hammons Pottinger, also had a summer kitchen, in Ohio, that was built in the 1940s; in the basement, it included a stove, a freezer, two large sinks, and a work table.

95. Marty Ahrens, *Home Candle Fires* (Quincy, MA: National Fire Protection Association, 2004).

96. Charles Lillie and Colin Mackenzie, *The British Perfumer*, 2nd ed. (London and New York: J. Souter and W. Seaman, 1822).

97. *Soap Making and Facts Interesting and Valuable to the Housekeeper and Others*, 24th ed. (Philadelphia: Penn Chemical Works, 1886); J. H. Chadwick and agent, *Economy Is Wealth: Make Your Own Soap with Lewis' Powdered Lye; Directions* (Boston: Lewis, 1881). That the Penn title is the 24th edition indicates that many such titles did not make it into the collections of libraries. For agricultural extension manuals, see, for example, *Crackling Soap, Boil Method* (Laramie, WY: Agricultural Extension Service, University of Wyoming, 1930).

98. Soapmaking instructions are still printed on cans of Red Devil lye. See also *Directions for the Use of Gillett's Lye in Soap-Making and Cleaning* [Modes d'Emploi de la Lessive Gillett pour Fabrication du Savon et Nettoyage] (Toronto: Gillett, 1940).

99. Lynn Alley, *Lost Arts* (Berkeley, CA: Ten Speed Press, 1995); Barbara J. Ciletti, *Making Great Cheese* (Asheville, NC: Lark Books, 1999); and Don Radke, *Cheese Making at Home* (Garden City, NY: Doubleday, 1974). Some of this recent literature addresses home production for market, such as Paul Kindstedt, *American Farmstead Cheese* (White River Junction, VT: Chelsea Green Publishing, 2005). Cheesemaking manuals have been published since the fifteenth century; see Irma Naso and Conflentia Panthaleon (fl. 1436–98), *Formaggi del Medioevo: La "Summa Lacticiniorum" di Pantaleone da Confienza* (Turin: Il Segnalibro, 1990); and Torquato Tasso and Bartholomew Dowe, *The Householders Philosophie: Anexed, A Dairie Booke* (1588; Amsterdam: Theatrum Orbis Terrarum, 1975).

100. Benjamin Thompson, Count Rumford, regarded the wasteful practice of watching fires burn in fireplaces, which he of course knew were much less fuel-efficient than stoves, as a disgrace to modern scientific life and expressed the hope that some device would be invented that could be watched in comfort that did not recklessly burn resources as fireplaces did. He might, therefore, have approved of television on energy-efficiency grounds; see his *Of Chimney Fire-Places*, 5th ed. (London: Cadell, 1798).

101. Ruth S. Cowan, "The Consumption Junction: A Proposal for Research Strategies in the Sociology of Technology," in *The Social Construction of Technological Systems*, ed. W. E. Bijker, T. P. Hughes, and T. J. Pinch. (Cambridge: MIT Press, 1987), 274. Some dining cars used charcoal grills; see *Ideas for Refreshment Rooms, Hotel, Restaurant, Lunch Room, Tea*

Room, Coffee Shop, Cafeteria, Dining Car, Industrial Plant, School, Club, Soda Fountain (Chicago: Hotel Monthly Press, 1923), 51–56.

102. Evidence for this tradition may be found in Molly Graham, *Cooking Out of Doors* (London: Andre Deutsch, 1960).

103. Elizabeth Von Arnim, *The Caravaners* (1930; London: Virago, 1989), 151.

104. Boy Scouts of America, *Boy Scouts Handbook: The First Edition, 1911* (Mineola, NY: Dover, 2005), 149–53, which includes instructions for washing the dishes afterwards.

105. *Cooking Out-of-Doors* (New York: Girl Scouts of the United States of America, 1946), 19–22.

106. *Oxford English Dictionary Online* (2000), s.v. "Barbecue," cites Washington Irving's *Knickerbocker* IV.IX (1849), 240; the noun referring to the rack or grill is much older, dating from the early eighteenth century. The verb form is attested in the seventeenth century as a cookery technique, but the practice had apparently not yet acquired its social and recreational elements.

107. *My Ride to the Barbecue* (New York: S. A. Rollo, 1860); and Robert M. Bradley, *A Sketch of Granny Short's Barbecue* (Louisville: Bradley & Gilbert, 1879).

108. Solomon W. Downs, *Speech . . . on the Compromise, Delivered on the 7th of June, 1851, at the Great Barbecue Given to Him Near Trinity, by the People of the Parishes of Catahoula and Concordia, Without Distinction of Party* (New Orleans: "True Delta" Book and Job Office, 1851); A. P. Eskridge, *The Citizens of Montgomery County Propose to Celebrate, on the 4th of July Next, at Christiansburg, the Anniversary of American Independence and the Passage of the Steam Cars of the Virginia Rail-Road through the Alleghany Mountain, by a Public Barbecue* (Montgomery County, VA: n.p., 1854); and Theodore G. Hunt, *Speech . . . at the Houma Barbecue, Parish of Terrebonne, on the 15th September* (New Orleans: n.p., 1855). See also Rosemary Brandau, *Early Fair Foods and Barbecuing*, Library Research Report Series RR-63 (Williamsburg VA: Colonial Williamsburg Foundation, 1984).

109. *Barbecue! and Pic-Nic! Camp of Ex-Confederate Veterans Will Hold a Grand Barbecue at Marlow, I.T. on Saturday, August 5, 1893* (Marlow, [MD?], 1893); and "Roasting Pigs at the Great Barbecue, G.A.R. Encampment," photograph, Louisville, KY, 1895, available from http://hdl.loc.gov/loc.pnp/cph.3b37598 (accessed 10 Sept. 2006).

110. Joe Flynn, *The Barbers' Barbecue: Comic Negro Song* (New York: Spaulding & Gray, 1895); E. C. Kammermeyer, *A Coonville Barbecue: Characteristic March and Two-Step* (Los Angeles: Eck Publishing, 1899); and William Kohnhorst, *An Old Kentucky Barbecue: March and Two-Step* (Louisville, KY: Finzer and Hamill, 1889). The religious context of African-American outdoor dining occasions is recalled in Honorée F. Jeffers, *The Gospel of Barbecue: Poems* (Kent, OH: Kent State University Press, 2000).

111. "An Elegant Barbacue [*sic*] Dinner," *Salem* [Mass.] *Gazette*, 3 June 1815, p. 2; and Augustus C. Taylor, *The Society of California Pioneers Thirty-First Anniversary Celebrated by a Grand Barbecue at Isabela Grove, Santa Cruz, Cal. September 9, 1881* (San Francisco: Society of California Pioneers, 1881).

112. Joseph T. Bonanno, *The Firehouse Grilling Cookbook* (New York: Broadway Books, 1998); and *The United Methodist Women of the Canterbury United Methodist Church of Mountain Brook, Alabama, Present the Fiesta Cookbook, Canterbury Cookout* (Mountain Brook, AL: Canterbury United Methodist Church, 1974).

113. "How to Build a Barbecue," *Sunset*, 10 June 1931; and *Sunset Barbecue Book*, rev. ed. (San Francisco: Lane, 1945). Note that the book is a second edition; I have not been able to verify the first.

114. *Ford Charcoal Briquets: Fuel of a Hundred Uses* (Dearborn, MI: Ford Motor Co., 1935); *Directions: Ford Grills* (Dearborn: Ford Motor Co., 1936); and *All Out for a Chick-n-Que: Cook Out Recipes Compliments of your Friendly Grocer* (Washington, DC: National Broiler Council, 1950).

115. Pamela Clark, *The Barbecue Cookbook* (Sydney: Australian Consolidated Press, 1987); Reuben Solomon, *Brilliant Barbecues* (Sydney: Murdoch Books, 1992); Mark Thomson, *Meat, Metal, and Fire* (New York: Harper Collins, 1999); De Waal Davis, *Braai Buddy* (Cape Town: Struik, 2003); and Kate Tunnicliff, Nigel Tunnicliff, and Tim Reeves, *Blistering Barbecues* (Bath: Absolute, 2004).

116. I am not making these up. *Big Daddy's Zubba Bubba BBQ* (New York: Roadside Amusements, 2005); Gideon Bosker, *Patio Daddy-O: 50s Recipes with a 90s Twist* (San Francisco: Chronicle Books, 1996); Bobby Flay and Julia Moskin, *Bobby Flay's Boy Gets Grill: 125 Reasons to Light Your Fire!* (New York: Scribner, 2004); Jeff Foxworthy and David Boyd, *The Redneck Grill* (Nashville, TN: Rutledge Hill Press, 2005); David Rick Joachim, *A Man, a Can, a Grill* (Emmaus, PA: Rodale, 2003); Rick Snider, *Secrets of Caveman Cooking for the Modern Caveman* (Phoenix, AZ: Golden West Publishers, 2001). For earlier examples in this tradition see *Old Hickory Barbecue Book: A Man's Guide to Outdoor Cooking* (St. Louis, MO: St. Louis Independent Packing Co., 1955); *Big Boy Barbecue Book* (New York: Grosset and Dunlap, 1960); James Beard, *Jim Beard's Argosy Barbecue Book* (New York: Popular Publications, 1952); William Patrick Magee and Ed Ainsworth, *Bill Magee's Western Barbecue Cookbook* (Culver City, CA: Murray & Gee, 1949). Counterexamples include Hyla Nelson O'Connor, *Today's Woman Barbecue Cook Book*, Fawcett Book no. 225 (New York: Arco Publishing, 1954); and Anita C. Dean and Robert E. Rust, *Outdoor Cookery for the Family*, Extension Bulletin 531 (Pullman, WA: Extension Service, Institute of Agricultural Sciences, State College of Washington, 1958). See also notes above on barbecuing for churches and on women as camp cooks.

117. Cecelia Ahern, *PS, I Love You* (New York: Hyperion, 2004), 209.

118. Jonathan Franzen, *The Corrections* (New York: Farrar, Straus and Giroux, 2001), 162.

119. Linda Majzlik, *Vegan Barbecues and Buffets* (Charlbury, Oxfordshire: Jon Carpenter, 1999); on grills see Richard Dennis, Judy Wallace, and Peter Calandruccio, *Pig Iron: Art Cookers; National Ornamental Metal Museum, April 28–July 7, 1996* (Memphis: The Museum, 1996).

120. On orreries, see, for example, James Ferguson, *The Use of a New Orrery* (London: The Author, 1746); on exhibited models in France see *Explication des Modeles des Machines et Forces Mouvantes, que l'on Expose à Paris dans la Rue de la Harpe, vis-à-vis Saint Cosme* (Paris, 1683). On educational models, see *Actien-Gesellschaft* (Darmstadt: Polytechisches Arbeits-Institut J. Schröder, 1889); and Robert Willis, *A System of Apparatus for the Use of Lecturers and Experimenters in Mechanical Philosophy* (London: J. Weale, 1851).

121. *Model of a Spanish Slaver* (Salisbury: W. Brodie and Co, 1840).

122. Thomas Savery, "An Account of Mr. Tho. Savery's Engine for Raising Water by the

Help of Fire," *Philosophical Transactions of the Royal Society* 21 (1699): 228; William Ray and Marly Ray, *The Art of Invention: Patent Models and Their Makers* (Princeton: Pyne Press, 1974); Royal Society of Arts, *A Catalogue of the Machines, Models, and Other Articles in the Repository of the Society* (London: R. Wilks for the Society, 1814); Commissioners of Patents, *Descriptive Catalogue of the Machines, Models, etc., in the Museum of the Commissioners of Patents at South Kensington*, 3rd ed. (London: George E. Eyre and William Spottiswoode, 1859); Eugene S. Ferguson and Christopher T. Baer, *Little Machines* (Greenville, DE: Hagley Museum, 1979).

123. Monsieur Le Quoy, *An Account of the Model in Relievo, of the Great and Magnificent City and Suburbs of Paris* (London: H. Hart, 1771); and Eduardo José de Moraes, *Plano Geral da Viação Ferrea da Provincia do Rio Grande do Sul: Memoria Apresentada à Consideração do Governo Imperial em 15 de Fevereiro de 1878* (Piranhas: Typographia da Locomotiva, 1882). See also *Description d'une Machine Mecanico-Tactique* (The Hague: H. Constapel, 1776).

124. William H. G. Kingston and W. J. Linton. 1861. *The Boy's Own Book of Boats* (London: S. Low, 1861); and *The Boys' Pump Book* (New York: Anson D. F. Randolph, 1860).

125. William Borman, *The Life of the Ingenious Agricultural Labourer, James Anderton the Founder and Builder of the Model of Lincoln Cathedral* (Newcastle-on-Tyne: J. Beall, 1868); Tyrrel E. Biddle, *A Treatise on the Construction, Rigging and Handling of Model Yachts, Ships and Steamers* (London: C. Wilson, 1879); and J. Du V. Grosvenor, *Model Yachts and Boats* (London: L. Upcott Gill, 1882).

126. "Men and Their Hobbies," *Scientific American* 62, no. 22 (1890): 343.

127. Paul N. Hasluck, *Building Model Boats*, "Work" Handbooks (London: Cassell, 1901).

128. Raymond F. Yates, *Model Making, Including Workshop Practice, Design, and Construction of Models*, 2nd ed. (New York: Norman W. Henley Publishing, 1925); Cyril Hall, *Model Making* (New York: R. F. Fenno, 1910); Percival Marshall, *Simple Electrical Working Models: How to Make and Use Them; A Practical Handbook for Electrical Amateurs and Students*, 13th ed., "Model Engineer" Series, no. 8 (London: P. Marshall, 1900); V. E. Johnson, *Modern Models: Including Full Instructions for Making and Using Model Aeroplanes, Dirigibles, Hydro-Aeroplanes, Mono-Rail Models, Wireless Telegraphy, X-ray Apparatus.* (London: C. Arthur Pearson, 1919); and Percival Marshall, *Machinery for Model Steamers* (London: P. Marshall, 1903). Many of these titles are in series, and a number of them were published in multiple editions.

129. "Some Famous 'Nefs,'" *Scientific American* 83, no. 15 (1900): 236; *The Naval and Marine Collection of the Late Lieut. Commander William Barrett R.N., London England* (New York: Anderson Galleries, 1926); and R. Morton Nance, *Sailing-Ship Models: A Selection from European and American Collections* (London: The Author, 1924). On the marketplace, see *Model Ship Supply Catalog, No. 6* (Mineola, NY: Model Ship Supply, [1920s]).

130. H. D. Jones, "Some Interesting Models," *Scientific American* 89, no 12 (1903): 200.

131. Christine Bose, Philip L. Bereano, and Mary Malloy, "Household Technology and the Social Construction of Housework," *Technology & Culture* 25, no. 1 (1984): 53–82.

132. Richard Neve, *Arts Improvement* (London: D. Brown, 1703); Thomas Powell, *The History of most Curious Manual Arts and Inventions very Pleasant and Profitable to all Lovers of Science and Recreation*, 3rd ed. (London: John Cole, 1675); Père Hulot, *L'Art du Tourneur Mécanicien* (Paris: M. Roubo, 1775).

133. Steven M. Gelber, *Hobbies* (New York: Columbia University Press, 1999), 155–92. On metalwork, see, for example, "Amateur Mechanics," *Scientific American* 43, no. 25 (1880): 390.

134. Note the differences between, for example, Charles and John Jacob Holtzapffel, *Turning and Mechanical Manipulation: Intended as a Work of General Reference and Practical Instruction, on the Lathe, and the Various Mechanical Pursuits Followed by Amateurs* (London: Holtzapffel & Co., 1843), with its appended essay by Charles Babbage, "Paper on the Principles of Tools, for Turning and Planing Metals"; and professional works like *The Handbook of Turning*, 2nd ed. (London: Saunders & Otley, 1846), and Peter Nicholson, *The Mechanic's Companion* (Philadelphia: James Locker, 1842). For hobbyists in *Scientific American*, see "Amateur: The Hand Lathe," *Scientific American* 13, no. 25 (1865): 388; "Having a Hobby," ibid. 32, no. 14 (1875): 216; "Value of a Hobby," ibid. 57, no. 26 (1887): 405; and review of J. H. Evans' *Ornamental Turning* (1903), ibid. 92, no. 2 (1905): 31.

135. Wilkie Collins, *The Law and the Lady* (1875; Oxford: Oxford University Press, 1992), 78.

136. Jeffrey Eugenides, *The Virgin Suicides* (New York: Warner Books, 1994), 27–28.

137. Sam Posey, *Playing with Trains* (New York: Random House, 2004). Jonathan Franzen, *The Corrections* (New York: Farrar, Straus and Giroux, 2001), 256, suggests that at least some retired railroad employees held model train enthusiasts in contempt; but it is in fact not unusual for such persons to participate in club layouts like the one at the Railroaders Memorial Museum in Altoona, PA.

138. William H. G. Kingston and W. J. Linton, *The Boy's Own Book of Boats* (London: S. Low, 1861).

139. Juvenile male aviation fantasies are described in detail in Fred Erisman, *Boys' Books, Boys' Dreams, and the Mystique of Flight* (Fort Worth: Texas Christian University Press, 2006). See also Bayla Singer, *Like Sex with Gods: An Unorthodox History of Flying* (College Station: Texas A&M University Press, 2003).

140. Artisans of both sexes wrote manuals on decorative leatherwork on both sides of the Atlantic. Among them were Charles G. Leland, *Leather Work*, 2nd ed. (London: Whittaker, 1901); George de Récy, *The Decoration of Leather* (London: Archibald Constable, 1905); Marguerite Charles, *Lessons in Leather Work* (New York: F. W. Devoe, 1906); Winifred H. Wilson and Marshall B. Willcox, *Leather Work* (New York: M. B. Willcox, 1908); *The Roycroft Leather-Book* (East Aurora, NY: Roycrofters, 1909); and Cécile Francis-Lewis, *A Practical Handbook on Leatherwork* (London: Francis-Lewis Studio, 1910).

141. William A. Maddox, *Historical Carvings in Leather* (San Antonio: Naylor, 1940); Leslie Spier, *Plains Indian Parfleche Designs*, University of Washington Publications in Anthropology 4, no. 3. (Seattle: University of Washington Press, 1931); Mary Trinick and Lilian E. Bristow, *A Portfolio of Designs for Leather Work Based on Historic Styles of Ornament* (London: Sir I. Pitman & Sons, 1932). For an example of a collector's guide, see Oliver Baker, *Black Jacks and Leather Bottells* (London: E. J. Burrow, 1921); for a museum catalogue see Günter Gall, *Deutsches Ledermuseum: Leder, Bucheinband, Lederschnitt, Handvergoldung, Lederwaren, Taschen*, 1st ed., Catalog vol. 1 (Offenbach am Main: Deutsches Ledermuseum, 1974).

142. Helen J. Biggart, *Leathercraft and Beading, Adapted for Camp Fire Girls* (New York: Camp Fire Outfitting Co., 1930); Lester Griswold, *Handbook of Craftwork in Leather, Horse-*

hair, Bead, Porcupine Quill and Feather, Indian (Navajo) Silver and Turquoise, 4th ed. (Colorado Springs: n.p., 1928); and Ann Macbeth, *Embroidered and Laced Leather Work* (London: Methuen, 1924).

143. Phyllis Hobson and Steven M. Edwards, *Tan Your Hide! Home Tanning Leathers and Furs* (Charlotte, VT: Garden Way Publishing, 1977); Larry Belitz, *Step-by-Step Brain Tanning the Sioux Way* (Hot Springs, SD: The Author, 1973); Mark Bowman, *It Takes Brains to Tan a Hide* (Ronan, MT: Dovetail Magazine, 1972); and Arlington C. Schaefer, *The Indian Art of Tanning Buckskin* (Roseburg, OR: Schaefer-Knudtson Publications, 1973). Note also the characteristically 1970s political agenda of David Runk, *Shoes for Free People* (Santa Cruz, CA: Unity Press, 1976).

144. See, for example, Robert Dossie, *The Handmaid to the Arts* (London: J. Nourse, 1758); and M. Bosc d'Antic, *Oeuvres de M. Bosc d'Antic: Contenant Plusieurs Mémoires sur l'Art de la Verrerie, sur la Faïencerie, la Poterie, l'Art des Forges, la Minéralogie, l'Électricité & sur la Médicine* (Paris, 1780). There was an enlarged second edition of Dossie's book by 1796. For the nineteenth century, see, for example, Simeon Shaw, *The Chemistry of the Several Natural and Artificial Heterogeneous Compounds Used in Manufacturing Porcelain, Glass, and Pottery* (London: Printed for the Author by W. Lewis and Son, 1837).

145. Some idea of the popularity of collecting and studying ceramics may be gained from the following nineteenth-century titles: *Catalogue of Old Dark Blue Crockery and Old China, Delft and other Ware* (Ayer, MA: Ayer Antique China Co., 1800); Alexandre Brongniart, *Traité des Arts Céramiques* (Paris: Béchet Jeune, 1844); Joan D'Huyvetter, *Objets Rares* (Ghent, Belgium: P. F. de Goesin-Verhaeghe, 1829); Henry Englefield and Henry Moses, *Ancient Vases from the Collection of Sir Henry Englefield, Bart* (London: H. G. Bohn, 1848); Joseph Marryat, *Collections towards a History of Pottery and Porcelain, in the 15th, 16th, 17th, and 18th Centuries* (London: J. Murray, 1850); George Richardson Porter, *A Treatise on the Origin, Progressive Improvement, and Present State of the Manufacture of Porcelain and Glass* (Philadelphia: U. Hunt, 1845); and Mrs. Bury Palliser, *The China Collector's Pocket Companion* (London: Sampson Low, Marston and Co., 1850). Mrs. Palliser also collected and wrote about lace and embroidery; her *Pocket Companion* had gone into many more editions by the end of the nineteenth century.

146. China painting was and is considered a primarily feminine craft, but the actual throwing and firing of ceramics is less heavily gendered in the twenty-first century. On this subject see *Pottery Painting for Amateurs* (London: E. Matthews & Sons, 1877); E. Campbell Hancock, *The Amateur Pottery and Glass Painter* (London: Chapman and Hall, 1879) and subsequent editions of this work, which was in the fourth edition by 1881; Hancock's 1880 *China Colours* (Worcester, UK: Hancock and Son; London: Reeves and Son); M. Louise McLaughlin, *China Painting: A Practical Manual for the Use of Amateurs in the Decoration of Hard Porcelain* (Cincinnati: R. Clarke, 1878), and her 1880 *Pottery Decoration under the Glaze* (Cincinnati: R. Clarke); John C. L. Sparkes, *A Handbook to the Practice of Pottery Painting*, 3rd ed. (London: Lechertier, Barbe, & Co. and Spottiswoode, 1879); Fred Miller, *Pottery-Painting* (London: Wyman & Sons, 1885); Adelaide Harriet Osgood, *How to Apply Royal Worcester, Matt, Bronze, La Croix and Dresden Colors to China: A Practical Elementary Hand-Book for Amateurs*, 5th rev. ed. (New York: Osgood Art School, 1891); *Illustrated Catalogue and Price List of White China and Materials for China Painting* (Philadelphia: Wright,

Tyndale & Van Roden, 1895); and Adelaide A. Robineau, Anna B. Leonard, and Felix Payant, eds., *Design* (Indianapolis: Saturday Evening Post Co., 1899–1930).

147. *Handbook of Modeling and Pottery Craft,* 5th ed. (Indianapolis: American Art Clay Co., 1936); *Hobby Craft Manual,* NavPers no. 15772 (Washington, DC: Navy Department, 1940); Walter A. De Sager, *Making Pottery: How to Do It,* 7th ser. (London: J. A. Divine and George Blachford, 1937); *Pottery Craft* (London: F. Warne; Peoria IL: Manual Arts Press, 1939); John Wolfe Dougherty, *Pottery Made Easy* (New York: Bruce Publishing, 1939); Ida W. Wheeler, *Playing with Clay* (New York: Macmillan, 1932); and Della F. Wilson, *Clay Modeling and Pottery* (Peoria, IL: Manual Arts Press, 1935).

148. Susanna W. Anthonies, *Pottery and Modelling* (London: Pittman & Sons, 1931); George J. Cox, *Pottery for Artists, Craftsmen, and Teachers* (New York: Macmillan, 1933) and the 1935 edition of this work; Bernard Lonsdale and Howard Ball, *Pottery: Coil Method* (Los Angeles: Division of Curriculum, Los Angeles County Schools, 1939); Dora Lunn, *Pottery in the Making* (Leicester, UK: Dryad Press, 1931), and the 1940 edition of this work (Peoria, IL: Manual Arts Press); Katherine V. Rogers, *A Handbook of Pottery* (Atlanta: National Youth Administration of Georgia, 1937); and two titles that associate pottery with scouting and the hedonization of Native American skills: *Pottery,* Merit Badge Pamphlet (New York: Boy Scouts of America, 1932), and Paul St. Gaudens, *Clay Craft,* Library of the Seven Crafts of the Camp Fire Girls, no. 7 (New York: Camp Fire Outfitting Company, 1931).

149. Edwine M. Winterbourne, "The Development of the Pleasure Crafters Motor Driven Pottery Wheel" (M.A. thesis, Colorado State College of Education, 1938).

150. For examples of nineteenth-century camping, see Paul Martin, "Popular Pastimes," sec. 49 in *Victorian Snapshots* (1939; New York: Arno Press, 1973).

151. For an amusing reference to horse-drawn RVing at the turn of the nineteenth century, see the novel by John Crowley, *Lord Byron's Novel: The Evening Land* (New York: William Morrow, 2005), 400.

152. Robert A. Carter, *Buffalo Bill Cody* (New York: J. Wiley, 2000), 160–68; Bayard Henry Paine, *The Famous Buffalo Hunt of Grand Duke Alexis of Russia* (n.p.: The Author, 1932); and William W. Tucker, *The Grand Duke Alexis in the United States of America* (1872; New York: Interland Publishing, 1972), 152–78.

153. On pot-hunting, see Clive Phillipps-Wolley, *Big Game Shooting: The Badminton Library of Sports and Pastimes* (London: Longmans, Green, 1894), 1:110; for the chamois, see 2:81. I am not an opponent of modern, licensed hunting, but the wholesale and wasteful slaughters of wildlife described in some nineteenth-century accounts of hunting have a repellent quality to the modern mind that their authors could not have imagined or intended.

154. Izaak Walton, *The Compleat Angler; or, The Contemplative Man's Recreation* (London: T. Maxey for Rich. Marriot, 1653), and dozens of later editions. The form of his title was actually quite common in the seventeenth century; see, for earlier examples than Walton's, Henry Peacham and Francis Delaram, *The Compleat Gentleman* (London: John Legat for Francis Constable, 1622); Edward Coke and Charles Calthrope, *The Compleat Copy-Holder, Wherein is Contained a Learned Discourse of the Antiquity and Nature of Mannors and Copy-Holds* (London: W. Lee and D. Pakeman, 1650); Thomas De Grey, *The Compleat Horseman and Expert Ferrier* (London: T. Harper, 1639); and John Roberts, *The Compleat Cannoniere* (London: J. Okes, 1639). A few of many later apparent imitators are H. Allen

Smith, *The Compleat Practical Joker* (Garden City, NY: Doubleday, 1953); Wilfrid Blunt and William T. Stern, *The Compleat Naturalist: A Life of Linnaeus* (New York: Viking, 1971); David A. Emery, *The Compleat Manager* (New York: McGraw-Hill, 1970); and two of many fishing titles, Guy Gilpatric, *The Compleat Goggler* (New York: Dodd Mead, 1957), and Norman Thelwell, *Thelwell's Compleat Tangler: Being a Pictorial Discourse of Anglers and Angling* (New York: E. P. Dutton, 1968).

155. On fishing as spiritual improvement, see Gervase Markham and Robert Venables, *The Pleasures of Princes; or, Good Men's Recreations* (1614; London: Cresset Press, 1927); John Lavicount Anderson, *The River Dove, with Some Quiet Thoughts on the Happy Practice of Angling* (London: W. Pickering, 1847); Leonard Hulit, *Fishing with a Boy: The Tale of a Rejuvenation* (Cincinnati: Steward Kidd, 1921); "S. J." and Nahum Tate, *The Innocent Epicure; or, The Art of Angling: A Poem* (London: S. Crouch by H. Playford and W. Brown, 1697); and Joseph Seccombe and Theodore Atkinson, *Business and Diversion Inoffensive to God, and Necessary for the Comfort and Support of Human Society* (Boston: S. Kneeland and T. Green, 1743). On fishing as a road to perdition, or at least to divorce, see Courtney Louise Borden, *Adventures in a Man's World: The Initiation of a Sportsman's Wife* (New York: Macmillan, 1933); Beatrice Cook, *Till Fish Us Do Part: The Confessions of a Fisherman's Wife* (New York: W. Morrow, 1949); Robert Traver, *Trout Madness* (New York: St. Martin's Press, 1960); Harold Tucker Webster, Edward Geary Zern, and Corey Ford, *To Hell with Fishing* (New York: D. Appleton-Century, 1945); Richard Penn, *Maxims and Hints for an Angler, and Miseries of Fishing* (London: J. Murray, 1839); and Lawrence Lariar, *Fish and Be Damned* (New York: Prentice-Hall, 1953). For parallels in knitting literature, see Stephanie Pearl-McPhee, *At Knit's End: Meditations for Women Who Knit Too Much* (North Adams, MA: Storey Publishing, 2005); the same author's *Yarn Harlot* (Kansas City, MO: Andrews McMeel Publishing, 2005); and Stitchy McYarnpants [pseud.], *The Museum of Kitschy Stitches* (Philadelphia, PA: Quirk Books, 2006).

156. American Veterinary Medical Association, *U.S. Pet Ownership and Demographics Sourcebook* (Schaumburg, IL: Membership & Field Services, American Veterinary Medical Association, 2002), 24–27; and Michael Korda, *Horse People* (New York: HarperCollins, 2003), 1, which puts the U.S. horse population at more than 13 million. I have used the more conservative of these two estimates. For the contrast with horses as prime movers, see Clay McShane and Joel A. Tarr, *The Horse in the City: Living Machines in the Nineteenth Century* (Baltimore: Johns Hopkins University Press, 2007).

157. Some agricultural fairs still have draft-horse competitions, but few of the competitors actually use horses for plowing and hauling in agricultural production. Horsepull .Com, "The Sport of Horsepulling," www.horsepulling.com/sport%20of%20horsepulling .htm (accessed 19 Mar. 2008).

158. American Horse Council, *Horse Industry Directory* (Washington, DC: American Horse Council in cooperation with American Horse Publications, 1999), 5.

159. Harry E. Cole and Louise P. Kellogg, *Stagecoach and Tavern Tales of the Old Northwest* (Carbondale: Southern Illinois University Press, 1997), 71–116, 197–209; Horace Sutton, *Travelers: The American Tourist from Stagecoach to Space Shuttle* (New York: Morrow, 1980), 34–36; Richard F. Palmer, *The "Old Line Mail": Stagecoach Days in Upstate New York* (Lakemont, NY: North Country Books, 1977), 36–48; Emily Williams and Helen Carda-

mone, *Stagecoach Country* (Turin, NY: Privately Printed, 1976), 3; Mary Einsel, *Stagecoach West to Kansas* (Boulder, CO: Pruett Publishing Co., 1970), 17–26; Ralph Moody, *Stagecoach West* (New York: T.Y. Crowell, 1967), 29–44; Cyril Noall and Daphne Du Maurier, *A History of Cornish Mail- and Stage-Coaches* (Truro, Eng.: D. B. Barton, 1963), 30–37, 58–62; and Oscar Osburn Winther, *Via Western Express and Stagecoach* (Stanford, CA: Stanford University Press, 1945), 24–27, 60–79, 115. See also the reference to seventeenth-century London vehicle drivers in Neil Hanson, *The Great Fire of London: in that Apocalyptic Year, 1666* (New York: Wiley, 2002), 31.

160. *The Delights of Coaching* (New York: Murphy, 1883); Anthony Burgess, *Coaching Days of England* (London: Elek, 1966); Stanley Harris, *The Coaching Age* (London: R. Bentley and Son, 1885); Martin E. Haworth, *Road Scrapings* (London: Tinsley Brothers, 1882); Reginald William Rives, *The Coaching Club* (New York: Privately Printed, 1935); Fairman Rogers, *A Manual of Coaching* (Philadelphia: J. B. Lippincott, 1900); Henfrey Smail and James Town, *Coaching Times and After: Including Some Old Coaching Celebrities, the Coaching Revival, Etc.* (Worthing, Sussex: Worthing Art Development Scheme by Aldridge Bros., 1948); W. Outram Tristam, Herbert Railton, and Hugh Thomson, *Coaching Days and Coaching Ways* (London: Macmillan, 1893); Edmund Vale and Thomas Hasker, *The Mail-Coach Men of the Late Eighteenth Century* (London: Cassell, 1960); Joseph Dommers Vehling, *Old Coaching Days and the White Horse Cellar, Piccadilly, Established A.D. 1720* (London: Hatchett's White Horse Cellars, 1920).

161. William M. Thackeray, *The Newcomes* (1854; Ann Arbor: University of Michigan Press, 1996), 138. For a coaching club in action, see Martin, "Society," sec. 37 in *Victorian Snapshots*. The modern parade float seems to have had a predecessor in the "coaching parade" of the turn of the twentieth century; see Allan Forbes and Ralph Mason Eastman, *Taverns and Stagecoaches of New England* (Boston: State Street Trust Co., 1953), 2:30–38.

162. Cyril Hall, *Models and How to Make Them* (London: C. Arthur Pearson, 1906).

163. Larry McMurtry, *The Colonel and Little Missie* (New York: Simon & Schuster, 2005), 5–6.

164. Michael Allen, *Rodeo Cowboys in the North American Imagination* (Reno: University of Nevada Press, 1998), 15–35, 193–214; Wayne S. Wooden and Gavin Ehringer, *Rodeo in America* (Lawrence, KS: University Press of Kansas, 1996), 7–16; Jeff Coplon, *Gold Buckle: The Grand Obsession of Rodeo Bull Riders* (San Francisco: HarperCollins West, 1995), 59–86, 87–104 (note chapter entitled "The Joy of Bucking"); and Dirk Johnson, *Biting the Dust* (New York: Simon & Schuster, 1994), 30–31. On bison, see Allison Fuss Mellis, *Riding Buffaloes and Broncos* (Norman: University of Oklahoma Press, 2003), 17–54.

165. The event of bull dogging was invented by African American rodeo star Bill Pickett in the early years of the twentieth century. Cecil Johnson, *Guts* (Fort Worth, TX: Summit Group, 1994).

166. Jane Austen, *Emma*, chap. 8.

167. Robert S. Baden-Powell, *Scouting for Boys* (1908; Oxford: Oxford University Press, 2004); and Randy Woo, "Sir Robert Baden-Powell: His Story," *Ultimate Boy Scouts of America History* Web site, http://users.aol.com/randywoo/bsahis/b-p.htm (accessed 17 Sept. 2006).

168. "About Us," *Girlguiding UK* Web site, 2006, www.girlguiding.org.uk/new/about (accessed 17 Sept. 2006).

169. Ernest T. Seton and Robert S. S. Baden-Powell, *Boy Scouts of America: A Handbook of Woodcraft, Scouting, and Life-craft* (New York: Doubleday Page, 1910).

170. "Juliette Gordon Low Biography, Founder of the Girl Scouts of the USA," *Girl Scouts of the USA* Web site, 2005, www.girlscouts.org/who_we_are/history/low_biography (accessed 17 Sept. 2006).

171. Ken Burns, Ric Burns, Geoffrey C. Ward, and David G. McCullough, *The Civil War*, Florentine Films, PBS Video, and WETA-TV (Alexandria, VA: PBS Video, 1989).

172. See, for example, Duncan B. Campbell and Brian Delf, *Greek and Roman Artillery, 399 BC–AD 363*, New Vanguard series, no. 89 (Oxford: Osprey, 2003), 41.

173. Jenny Thompson, *War Games* (Washington DC: Smithsonian Books, 2004), xiii.

174. Geoffrey C. Ward et al., *The Civil War: An Illustrated History* (New York: Knopf, 1990), 413–14.

175. James Oscar Farmer, "Playing Rebels: Reenactment as Nostalgia and Defense of the Confederacy in the Battle of Aiken," *Southern Cultures* 11, no. 1 (2005): 46–73.

176. Barbara Brackman, *Civil War Women: Their Quilts, Their Roles, Activities for Re-enactors* (Lafayette, CA: C&T Publishing, 2000); and Maggie Black, *Food and Cooking in Medieval Britain: History and Recipes* (England: Historic Buildings & Monument Commission, 1985).

177. There are fictional references to the cathartic qualities of paintball in Christopher A. Bohjalian, *The Double Bind* (New York: Shaye Areheart Books, 2007), 52 and 121.

178. Seton and Baden-Powell, *Boys Scouts of America*, 163.

179. James William Gibson, *Warrior Dreams* (New York: Hill and Wang, 1994), 121–41.

180. Kristen Haring, "The 'Freer Men' of Ham Radio: How a Technical Hobby Provided Social and Spatial Distance," *Technology and Culture* 44, no. 4 (2003): 734–61; and Claude S. Fischer, "'Touch Someone': The Telephone Industry Discovers Sociability," *Technology and Culture* 29, no. 1 (1988): 32–61.

181. Haring, "'Freer Men,'" 749; Keir Keightley, "'Turn It Down!' She Shrieked: Gender, Domestic Space, and High Fidelity, 1948–1959," *Popular Music* 15, no. 2 (1996): 149–77. See also the humorous fictional reference to audiophiles in Franzen, *The Corrections*, 406. There are similar references to gender tensions in electronic music in Trevor J. Pinch and Frank Trocco, *Analog Days: The Invention and Impact of the Moog Synthesizer* (Cambridge: Harvard University Press, 2002), 78, 155–70.

182. Hugh G. J. Aitken, *The Continuous Wave* (Princeton: Princeton University Press, 1985), 144, 458, 470, 472–73, 512.

183. Susan J. Douglas, *Inventing American Broadcasting, 1899–1922* (Baltimore: Johns Hopkins University Press, 1989), 191–92. For the rhetoric of pleasure in ham radio, see ibid., 187, 206, 297, 301; and Aitken, *Continuous Wave*, 473.

184. Jesse Walker, *Rebels on the Air* (New York: New York University Press, 2001), 13–28; American Radio Relay League, *The Beginner's Guide to Amateur Radio* (Englewood Cliffs, NJ: Prentice–Hall, 1982); and Richard D. Kuslan and Louis I. Kuslan, *Ham Radio* (Englewood Cliffs, NJ: Prentice–Hall, 1981).

185. Edward T. Canby, *Home Music Systems*, rev. ed. (New York: Harper, 1955), 4.

186. Eric W. Rothenbuhler and John D. Peters, "Defining Phonography: An Experi-

ment in Theory," *Musical Quarterly* 81, no. 2 (1997): 253; and Norman Eisenberg, *Hi-Fi* (New York: Random House, 1958), 11.

187. Joseph O'Connell, "The Fine Tuning of a Golden Ear: High-End Audio and the Evolutionary Model of Technology," *Technology and Culture* 33, no. 1 (1992): 1–37.

188. A useful source for disambiguating readership of photography titles is Robert S. Sennett, *The Nineteenth-Century Photographic Press* (New York: Garland Publishing, 1987). Sennett identifies as significant contributions to the amateur literature Robert Hunt, *A Popular Treatise on the Art of Photography* (1841; Athens: Ohio University Press, 1973); A. Bisbee, *The History and Practice of Daguerreotyping* (1853; New York: Arno Press, 1973); George B. Coale, *Manual of Photography, Adapted to Amateur Practice* (Philadelphia: J. B. Lippincott, 1858); Lake Price, *A Manual of Photographic Manipulation* (1868; New York: Arno Press, 1973); Edward John Wall, *A Dictionary of Photography for the Amateur and Professional Photographer* (New York: E. & H. T. Anthony, 1889); and Thomas Cradock Hepworth, *The Book of the Lantern* (New York: E. L. Wilson, 1889).

189. Grace Seiberling and Carolyn Bloore, *Amateurs, Photography, and the Mid-Victorian Imagination* (Chicago: University of Chicago Press, in cooperation with the International Museum of Photography, 1986), 4–116.

190. Robert Johnston, "Slanguage of American Photographers," *American Speech* 15, no. 4 (1940): 357.

191. George Gilbert, *Collecting Photographica* (New York: Hawthorn Books, 1976); and Floyd Rinhart and Marion Rinhart, *The American Daguerreotype* (Athens: University of Georgia Press, 1981).

192. Reese Jenkins, *Images and Enterprise* (Baltimore: Johns Hopkins University Press, 1975), 172–87.

193. See, for example, the experiments in aerial photography and ballooning in Félix Nadar, *Le Droit au Vol*, 2nd ed. (Paris: J. Hetzel, 1865); and the discussion of scientific experiments in Martin, *Victorian Snapshots*, 12–13.

194. See, for example, Edward Augustus Samuels, *With Fly-Rod and Camera* (New York: Forest and Stream Publishing, 1890).

195. "An Instantaneous Shutter for Hand Cameras," *Scientific American* 63, no. 24 (1890): 376.

196. "Photographs Taken by Magic," *Scientific American* 81, no. 10 (1899): 150; and Percy Collins, "Table-Top Photography," *Scientific American* 99, no. 14 (1908): 224.

197. Barbara Kingsolver, *Animal Dreams* (New York: Harper Perennial, 1991), 69–70.

198. Robert C. May, *The Lexington Camera Club, 1936–1972* (Lexington: University of Kentucky Art Museum, 1989). Judging by the photograph on the last page before the Notes section of this unpaginated title, this particular club was an all-male, and of course all-white, preserve.

Chapter 5 · Why, When, and How Do Technologies Hedonize?

1. A. H. Crosfield, *A Plea for the Eight-Hours Day in Continuous Processes* (London: International Association for Labor Legislation, 1912); National Industrial Conference Board, *Practical Experience with the Work Week of Forty-Eight Hours or Less* (New York: National In-

dustrial Conference Board, 1921); Roy Rosenzweig, *Eight Hours for What We Will* (Cambridge: Cambridge University Press, 1983).

2. For the concept of the double day, see Charlene Gannagé, *Double Day, Double Bind: Women Garment Workers* (Toronto, Ont.: Women's Press, 1986).

3. For a humorous view of the uselessness of parlors, see Anthony Trollope, *Rachel Ray* (New York: Harper, 1864), 8–9.

4. My late father-in-law, Marion Pottinger, a Cincinnati millwright, had four separate hobby spaces associated with a two-bedroom former farmhouse: basement, garage, chicken coop, and a 60 × 80 foot hobby garden. His wife, Alta Opal Hammons Pottinger, had a sewing room in a small former parlor at the front of the house.

5. U.S. Bureau of the Census, "Retail Trade: Appendixes," in *United States Census of Business, 1948* (Washington: U.S. Government Printing Office, 1951), 2: 25.07, where hobbies and needlework are aggregated with other retailing.

6. Eleanor Chandler, "A Short History of the Founding and Early Years of the Embroiderers' Guild (American Branch) Inc.," *Needle Arts* 1, no. 1 (1970): 7–15.

7. This is a more controversial point than one might suppose. Definitions of "hobby" vary considerably. For example, the Craft and Hobby Association, the international hobby industry's organization, does not consider gardening a hobby because its membership does not include the home-gardening industry, which, of course, has its own organization, the Lawn and Garden Marketing and Distribution Association. Harris Interactive and American Demographics also study American leisure activities and rank them by popularity, but neither distinguishes between plain sewing and needlework as a hobby craft.

8. Mariska Karasz, *Adventures in Stitches* (New York: Funk & Wagnalls, 1949); and *Open Chain* (Menlo Park: Calif., Fibar Designs, 1979).

9. Jacqueline Enthoven, *The Stitches of Creative Embroidery* (New York: Van Nostrand Reinhold, 1965); Enthoven, *Stitchery for Children* (New York: Reinhold, 1968); Nik Krevitsky, *Batik Art and Craft* (New York: Reinhold, 1964); Krevitsky, *Stitchery: Art and Craft* (New York: Reinhold, 1966); and Nik Krevitsky and Lois Ericson, *Shaped Weaving* (New York: Van Nostrand Reinhold, 1974).

10. The works of these authors are too numerous to cite them all; a few examples include the following.

American authors. By Erica Wilson: *Crewel Embroidery* (New York: Scribner, 1962); *Fun with Crewel Embroidery* (New York: Scribner, 1965); *The Craft of Crewel Embroidery* (New York: Scribner, 1971); *Erica Wilson's Embroidery Book* (New York: Scribner, 1973); *16 Needlepoint Designs from the New World of Plastic Canvas* (New York: Newspaperbooks, 1977); *The Animal Kingdom of Erica Wilson* (New York: Newspaperbooks, 1977); and *Ask Erica* (New York: Scribner, 1977). By Mildred J. Davis: *The Art of Crewel Embroidery* (New York: Crown Publishers, 1962); *Early American Embroidery Designs* (New York: Crown Publishers, 1969); *Embroidery Designs, 1780–1820* (New York: Crown Publishers, 1971); and *The Dowell-Simpson Sampler* (Richmond: Textile Resource and Research Center, Valentine Museum, 1975). By Muriel L. Baker: *A Handbook of American Crewel Embroidery* (Rutland, VT: C. E. Tuttle, 1966); *The ABC's of Canvas Embroidery* (Sturbridge, MA: Old Sturbridge Village, 1968); *The XYZ's of Canvas Embroidery* (Sturbridge, MA: Old Sturbridge Village, 1971); *Stumpwork* (New York: Scribner, 1978); *The Scribner Book of Embroidery Designs* (New York:

Scribner, 1979); and, with Margaret Lunt, *Blue and White: The Cotton Embroideries of Rural China* (New York: Scribner, 1977). By Bucky King: *Creative Canvas Embroidery* (New York: Hearthside Press, 1963); and Bucky King and Jude Martin, *Ecclesiastical Crafts* (New York: Van Nostrand Reinhold, 1978). By Elsa S. Williams, *Bargello* (New York: Van Nostrand Reinhold, 1967); *Heritage Embroidery* (New York: Reinhold Publishing, 1967); *Creative Canvas Work* (New York: Van Nostrand Reinhold, 1971); and *The Joy of Stitching* (New York: Van Nostrand Reinhold, 1978).

British authors. By Constance Howard: *Design for Embroidery* (London: Batsford, 1956); *Inspiration for Embroidery*, 2nd ed. (London: Batsford, 1967); *Embroidery and Colour* (New York: Van Nostrand Reinhold, 1976); *Textile Crafts* (New York: Scribner, 1978); and *Constance Howard's Book of Stitches* (London: Batsford, 1979). By Diana Springall: *Canvas Embroidery* (London and Newton Center, MA: Batsford and Branford, 1969). By Mary Gostelow: *A World of Embroidery* (London: Mills & Boon, 1975); *Blackwork* (London: Batsford, 1976); *Embroidery South Africa* (London: Mills and Boon, 1976); *Embroidery of All Russia* (New York: Scribner, 1977); *Art of Embroidery* (London: Weidenfeld and Nicolson, 1979); *Mary Gostelow's Embroidery Book* (New York: Dutton, 1979); and Mary Gostelow and Susannah Read, *The Complete International Book of Embroidery* (New York: Simon and Schuster, 1977). By Beryl Dean: *Ecclesiastical Embroidery* (London: Batsford, 1958); *Church Needlework* (London: Batsford, 1961); *Ideas for Church Embroidery* (Newton Center, MA: Branford, 1968).

11. Sherlee Lantz and Maggie Lane, *A Pageant of Pattern for Needlepoint Canvas* (New York: Atheneum, 1973), and Sherlee Lantz, *Trianglepoint* (New York: Viking, 1976). As of 23 September 2006, 569 libraries listed in WorldCat held *A Pageant of Pattern* and 453 held *Trianglepoint*.

12. Peter S. Beagle and Baron Wolman, *American Denim* (New York: H. N. Abrams, 1975); and Alexandra Jacopetti and Jerry Wainwright, *Native Funk and Flash* (San Francisco: Scrimshaw Press, 1974).

13. Trevor J. Pinch and Frank Trocco, *Analog Days: The Invention and Impact of the Moog Synthesizer* (Cambridge: Harvard University Press, 2002), 89–108; and Anna Thomas, *The Vegetarian Epicure* (New York: Vintage Books, 1972), 9.

14. Mark Dittrick, *Hard Crochet* (New York: Hawthorn Books, 1978).

15. See, for example, Nedda C. Anders and Bucky King, *Appliqué, Old and New* (New York: Hearthside Press, 1967); Beryl Dean, *Creative Appliqué* (London and New York: Studio Vista and Watson-Guptill, 1970); and American Craftsmen's Council, *Fabric Collage* (New York: Museum of Contemporary Crafts, 1965).

16. Carrie A. Hall and Rose G. Kretsinger, *The Romance of the Patchwork Quilt in America* (Caldwell, ID: Caxton Printers, 1936); Florence Peto, *American Quilts and Coverlets* (New York: Chanticleer Press, 1949); Marguerite Ickis, *The Standard Book of Quilt Making and Collecting* (New York: Dover, 1959); and Ruby Short McKim, *One Hundred and One Patchwork Patterns*, rev. ed. (New York: Dover, 1962).

17. Jonathan Holstein, *Abstract Design in American Quilts* (New York: Whitney Museum of American Art, 1971); see also his *The Pieced Quilt* (Greenwich, CN: New York Graphic Society, 1973). An exhibition of quilts at the Baltimore Museum of Art was nearly contemporaneous with the Whitney show but did not receive as much attention at the

time: Dena S. Katzenberg, *The Great American Cover-Up* (Baltimore: Baltimore Museum of Art, 1971). For an earlier example, see Margaret E. White, *Quilts and Counterpanes in the Newark Museum* (Newark: Newark Museum, 1948).

18. Marilynn J. Bordes, *Twelve Great Quilts from the American Wing: Catalogue* (New York: Metropolitan Museum of Art, 1974); Lincoln Quilters Guild, *Quilts from Nebraska Collections* (Lincoln: The Gallery, 1974); and University of Kansas Museum of Art, *150 Years of American Quilts* (Lawrence: The Museum, 1973).

19. Cuesta Benberry, "Quilts of the Late Victorian Era," *Nimble Needle Treasures* 5, no. 3 (1973): 2–3; Lenice Ingram Bacon, *American Patchwork Quilts* (New York: Morrow, 1973); Carleton L. Safford and Robert C. Bishop, *America's Quilts and Coverlets* (New York: Dutton, 1972); Patsy Orlofsky and Myron Orlofsky, *Quilts in America* (New York: McGraw-Hill, 1974); and Mary Washington Clarke, *Kentucky Quilts and Their Makers* (Lexington: University Press of Kentucky, 1976).

20. Pattie Chase and Mimi Dolbier, *The Contemporary Quilt* (New York: Dutton, 1978); Joanne Mattera, *The Quiltmaker's Art* (Asheville, NC: Lark Books, 1982); Kay Parker, *Contemporary Quilts* (Trumansburg, NY: Crossing Press, 1981); Quilt National, *The New American Quilt* (Asheville, NC, and New York: Lark Communications and Hastings House Publishers, 1981); and Charlotte Robinson, *The Artist and the Quilt* (New York: Knopf, distributed by Random House, 1983). For an idea of how this aesthetic trend has developed, see Dianne S. Hires and Renegades (quilting group), *Oxymorons: Absurdly Logical Quilts!* (Paducah, KY: American Quilter's Society, 2001).

21. Paula Mariedaughter, "A Brief History of the Art Quilt," [c. 2000], www.quilt professionals.com/quiltdirectory/artquilthistory.htm (accessed 30 Sept. 2006). For a more analytical discussion of this movement, see Thalia Gouma-Peterson and Patricia Mathews, "The Feminist Critique of Art History," *Art Bulletin* 69, no. 3 (1987): 326–57.

22. Patricia Mainardi, "Quilts: The Great American Art," *Feminist Art Journal* (Winter 1973) reprinted in Norma Broude and Mary D. Garrard, *Feminism and Art History* (New York: Harper & Row, 1982).

23. The negative qualities of Chicago's personality were considerably mitigated by the sympathetic and sensitive management of her chief textile artist, Susan Hill. For illustrations and discussion of these works, see Judy Chicago, *The Dinner Party* (Garden City, NY: Anchor Press/Doubleday, 1979), and her *Birth Project* (Garden City, NY: Doubleday, 1985); Judy Chicago and Susan Hill, *Embroidering Our Heritage* (Garden City, NY: Anchor Books/ Doubleday, 1980); and Amelia Jones and Laura Cottingham, *Sexual Politics: Judy Chicago's Dinner Party in Feminist Art History* (Los Angeles: UCLA at the Armand Hammer Museum of Art and Cultural Center in association with University of California Press Berkeley, 1996).

24. Susan S. Jorgensen and Susan S. Izard, *Knitting into the Mystery* (New York: Morehouse Publishing, 2003); Susan Gordon Lydon, *The Knitting Sutra* (San Francisco: Harper San Francisco, 1997); Tara Jon Manning, *Mindful Knitting* (Boston: Tuttle Publishing, 2004); Bernadette Murphy, *Zen and the Art of Knitting* (Avon, MA: Adams Media Corp., 2002); Linda Roghaar and Molly Wolf, *KnitLit* (New York: Three Rivers Press, 2002); the same authors' *KnitLit Too* (New York: Three Rivers Press, 2004); Afi Scruggs, *Beyond Stitch and Bitch* (Hillsboro, OR: Beyond Words Publishing, 2004); Debbie Stoller, *Stitch 'n Bitch*

(New York: Workman, 2003); Roberta Mintz Levine, "Unplugged: Yarning Curve; Knitting Warms Up Kids for Other Skills at Local School," *Pittsburgh Magazine*, Oct. 2005, 74; Stephanie Piro, "When Tom Left, I Took Up Knitting" [cartoon], *Chronicle of Higher Education*, 13 Jan. 2006, B18; Susan Banks, "Story of How Knitter Became an Author Is Quite a Yarn," *Pittsburgh Post-Gazette*, 5 Apr. 2006; Nora Isaacs, "The Art of Failure: Learning to Knit Didn't Seem That Hard, But as My Scarf Unraveled, So Did I," *Alternative Medicine*, Nov./Dec. 2005, 115–16; and Herbert Benson and Julie Corliss, "Ways to Calm Your Mind," *Newsweek*, 27 Sept. 2004, 47.

25. Manning, *Mindful Knitting:* 4-6; Sam Posey, *Playing with Trains* (New York: Random House, 2004), 81–89; and John Gierach, *Still Life with Brook Trout* (New York: Simon & Schuster, 2005), 15–60.

26. There is a fictional reference to the love of old tools in Arturo Pérez-Reverte, *The Nautical Chart*, trans. M. S. Peden (New York: Harcourt, 2001), 30.

27. Carroll E. Shaw, "Supervision of Teen-Age Rocketeers," *National Fire Protection Association Quarterly* 52, no. 2 (1958): 99–103; and John Bridgewater and James S. Mountain, *Rocket Amateur's Guide-Book* (New York: Space Products Corp., 1958).

28. James B. Twitchell and Kenneth C. Ross, *Where Men Hide* (New York: Columbia University Press, 2006), 67–80, 93–104, 183–97.

29. *At Our Leisure: A Complete Picture of How Americans Spend Their Free Time* (Long Island City, NY: Alert Publishing, 1992), 191.

30. Donald MacKenzie, "Marx and the Machine," *Technology and Culture* 25, no. 3 (1984): 500.

31. For an example of this kind of market adaptation, see Posey, *Playing with Trains*, 22.

32. Yuzo Takahashi, "A Network of Tinkerers: The Advent of the Radio and Television Receiver Industry in Japan," *Technology and Culture* 41, no. 3 (2000): 484.

33. *The 2002 Entertainment, Cultural, and Leisure Market Research Handbook* (Norcross, GA: Richard K. Miller & Associates, 2002), 396; *Gardening—U.S.* (London: Mintel International Group, 2003), and the same firm's *Outdoor Barbecue—U.S.* (London: Mintel International Group, 2005).

34. Jill Fruchter and Michael Schau, *At Our Leisure*, 3rd ed. (New York: EPM Communications, 1997), 252.

35. *At Our Leisure* (1992), 190–94.

36. For a theory of animal play, see Gordon M. Burghardt, *The Genesis of Animal Play: Testing the Limits* (Cambridge: MIT Press, 2005).

37. Jacques Ellul, *The Technological Society* (New York: Vintage Books, 1964); Theodore Roszak, *Where the Wasteland Ends* (Garden City, NY: Doubleday, 1972); René J. Dubos, *So Human an Animal* (New York: Scribner, 1968); Langdon Winner, *Autonomous Technology* (Cambridge: MIT Press, 1977). See also the discussion of these authors' arguments in Samuel C. Florman, "Antitechnology," pp. 45–46 in *The Existential Pleasures of Engineering*, 2nd ed. (New York: St. Martin's Press, 1994).

38. Albert Borgmann, *Technology and the Character of Contemporary Life* (Chicago: University of Chicago Press, 1984), 201–15.

39. Don Ihde, *Philosophy of Technology* (New York: Paragon House, 1993), 69.

40. See, for example, Ellul, *Technological Society*, 400–402; and Steven M. Gelber *Hob-*

bies: Leisure and the Culture of Work in America (New York: Columbia University Press, 1999), 295–99. On television, see Albert Borgmann, "The Moral Assessment of Technology," pp. 211–13 in *Democracy in a Technological Society,* ed. L. Winner (Dordrecht: Kluwer, 1992).

41. Ronald M. Radano, "Interpreting Muzak: Speculations on Musical Experiences in Everyday Life," *American Music* 7, no. 4 (1989): 458.

Appendix A · *Biases of Collecting and Connoisseurship*

1. See, for example, Marcus B. Huish, *Samplers and Tapestry Embroideries,* 2nd ed. (New York: Dover, 1970), pl. III and fig. 52.

2. Works by these authors include Mildred J. Davis, *Early American Embroidery Designs* (New York: Crown Publishers, 1969); Ethel Stanwood Bolton and Eva J. Coe, *American Samplers* (Boston: Massachusetts Society of the Colonial Dames of America, 1921); Betty Ring, *Needlework: An Historical Survey,* new and expanded ed. (Pittstown, NJ: Main Street Press, 1984); and Susan B. Swan, *Plain and Fancy* (New York: Holt, Rinehart and Winston, 1977). In Ring's book, see especially her remarks on pp. 81, 87, and 183, and those of Larry Salmon on p. 177.

3. For example, there are only brief references to crochet, knitting or tatting in Margaret B. Schiffer, *Historical Needlework of Pennsylvania* (New York: Scribner, 1968); Swan, *Plain and Fancy*; and Georgiana B. Harbeson, *American Needlework* (New York: Bonanza Books, 1938).

4. Huish, *Samplers,* 10; Averil Colby, *Samplers* (London: B. T. Batsford, 1964), 19.

5. Rachel Maines, "Evolution of the Potholder: From Technology to Popular Art," *Journal of Popular Culture* 19, no. 1 (1985): 3-34.

6. A. Rupert Hall, "Early Modern Technology to 1600," in *Technology in Western Civilization,* ed. M. Kranzberg and C. W. Pursell (New York: Oxford University Press, 1967), 96.

7. Steven M. Gelber, *Hobbies: Leisure and the Culture of Work in America* (New York: Columbia University Press, 1999), 55–77.

8. History curators are less inclined to these forms of snobbery than are art curators.

9. See, for example, Marianne M. C. Alford, and Elizabeth G. Holahan, *Needlework as Art* (London: S. Low, Marston, Searle, and Rivington, 1886).

10. Huish, *Samplers,* 120. See also p. 52 for Huish's views on the deterioration of design quality.

11. Candace Wheeler, *The Development of Embroidery in America* (New York: Harper & Brothers, 1921): 96.

12. Martha Stearns, *Homespun and Blue* (New York: Scribner, 1963).

13. Davis, *Early American Embroidery Designs,* 151.

14. Swan, *Plain and Fancy,* 6. Swan and a co-author had commented on this supposed trend in an earlier work, Mary T. Landon and Susan B. Swan, *American Crewelwork* (New York: Macmillan, 1970), 87. See also *Plain and Fancy,* 88, 109, 159.

15. This expression is customary in the world of art and artifacts, but as I object to periodizing events in the American republic by reference to a British monarch, I do not use

it in my discussions of nineteenth-century needlework except when quoting other authors who use the term.

16. Garrel S. Pottinger, personal communication, May 2006.

17. Gelber, *Hobbies*, 161–62, 167, 171. Whitework is white-on-white embroidery, a type of fancywork; see Thérèse de Dillmont, *Encyclopedia of Needlework*, 2nd ed. (1880s; Philadelphia: Running Press, 1978), 47–84; Harbeson, *American Needlework*, 73–75; Swan, *Plain and Fancy*, 132; and Jane Dew and Vivo Watkins, *Whitework* (London and New York: C. Lets, distributed by Sterling Publishing, 1990). Shell work is a method of decorating containers with putty and shells, not a textile technique; see Hannah Robertson, *The Young Ladies' School of Arts*, 2nd ed. (Edinburgh: W. Rudiment Junior, 1767); and Levine B. Urbane, *Art Recreations* (Boston: J. E. Tilton, 1864). Gelber also seems to believe that embroidery canvases were printed with colored patterns as early as 1850 (p. 166) and is apparently unaware that patterns were hand-painted on fabrics for embroiderers for many centuries before this supposed "bastardization" of traditional feminine arts. For Swan's views of Berlinwork as an inferior type of needlework, see *Plain and Fancy*, 206. On painting by numbers, see William L. Bird, *Paint by Number* (Washington, DC, and New York: Smithsonian Institution and National Museum of American History in association with Princeton Architectural Press, 2001.)

Appendix B · Methodological Notes

1. For a brief history of the Online Computer Library Center, see Scott Carlson, "Frederick G. Kilgour, Developer of Famed Electronic-Library Catalog, Is Dead," *Chronicle of Higher Education* 52, no. 40 (2006): A29.

2. Matheo Pagano, *Giardineto Nouo di Punti Tagliati et Gropposi: Per Exercito & Ornamento delle Donne* (Venice: the Author, 1550).

3. My comments should not be construed as a criticism of these rules and conventions, which taken as a whole, are a marvel of stubbornly rational efforts, over more than a century, to make our global intellectual heritage accessible to researchers and to create comprehensible paths through the glorious chaos that is our legacy of texts and images. See, for example, *Anglo-American Cataloging Rules*, 2nd ed. (Chicago: American Library Association and Library of Congress, Descriptive Cataloging Division, 1998).

4. Devon D. Brewer and John M. Roberts, "Measures and Tests of Heaping in Discrete Quantitative Distributions," *Journal of Applied Statistics* 28, no. 7 (2001): 887–96.

5. Library of Congress Copyright Office, *Copyright Registration for Serials,* Circular No. 62 (Washington, DC: Library of Congress, January 2004), 1. This title, of course, is itself a biennial serial.

6. Jeffrey Y. Young, "Library Groups Plan a Merger That Will Fuse Two Catalogs," *Chronicle of Higher Education* 52, no. 37 (2006): A36.

7. Cesare Vecellio, *Pattern Book of Renaissance Lace: A Reprint of the 1617 Edition of the "Corona delle Nobili et Virtuose Donne,"* ed. Stanley Appelbaum and Mary Carolyn Waldrep (New York: Dover, 1988).

8. Juan de Alcega, *Tailor's Pattern Book, 1589: Facsimile,* ed. J. L. Nevinson (New York: Costume & Fashion Press, 1999).

Except as noted, definitions in quotation marks are from S. F. A. Caulfeild and Blanche A. Saward, *Dictionary of Needlework,* 2nd ed. (London: A. W. Cowan, 1887).

Aniline dye. The first synthetic dye, based on coal tar, introduced in the mid-nineteenth century.

Antimacassar. Textile attached to the upper back of a chair or sofa to protect it from hair (macassar) oil.

Appliqué. "A French term, signifying the sewing of one textile over another."

Aran knitting. A type of ornate, raised knitting, typically with groups of stitches crossed over each other to form cables, in heavy wool yarn, thought to have originated in the Aran Islands in Galway Bay on the western coast of Ireland.

Assisi work. "A form of counted-thread embroidery based on an ancient Italian tradition where the background is filled with embroidery stitches and the main motifs are left void, i.e., unstitched . . . Cross-stitch is used for the background, and Blackwork embroidery . . . is then used to outline the motif and create the surrounding decorative scroll-work." (Wikipedia)

Audiophilia. The love of high-fidelity recorded music and other sound.

Backstitch. "In making a running [line of stitches], a stitch is taken back into the material beyond where the thread was last drawn through."

Basting. "Otherwise called Tacking.—Derived from the old German *bastan,* to sew, or *besten,* to bind. This term is chiefly employed by tailors, while Tacking is used by women. The term is used to signify the light runnings made by taking up a stitch at long distances successively, to keep the separate portions of a garment or other article in position, preparatory to their being sewn together."

Berlinwork. Counted embroidery worked in wool over Berlin canvas, in which "every two strands . . . are drawn together, thus forming squares, and leaving open spaces for the wool, with which it may be embroidered. It is more easily counted and worked than the ordinary sorts, and is a great improvement upon the old Penelope canvas, the threads of which were woven in equal distance throughout, taking, of course, much more time to count and separate them. It may be procured in almost all widths and all degrees of fineness."

Blackwork. Sometimes called *Spanish work,* a counted thread technique in black on even-weave fabrics.

Bride. "The connecting threads thrown across spaces in all Needle-point Laces, whether imitation or real, and known as Brides, Bride Claires, Coxcombs, Pearls, Legs and Ties. These threads are arranged so that they connect the various solid parts of the lace together." Also called *bars.*

Candlewicking. A type of raised embroidery, usually on white fabric, using the heavy, loosely spun cotton or linen yarn called *candlewick,* popular during the Colonial Revival in the United States.

Canvaswork. Embroidery worked over a loosely woven fabric in which the number of warp and weft ends per inch are the same.

Chain stitch. "A stitch used in Embroidery, Tambour Work, and Crochet. The manner of working it for embroidery is as follows: Bring the needle, threaded, from the back of the material, and form a loop on the right side, and keep this loop steady with the left thumb, return the needle close to where it came out, bring the needle up again in the centre of the loop, and pull the thread evenly up; then form another loop and return the needle as before."

Chenille. "The French for 'caterpillar.' A beautiful description of cord employed for embroidery and decorative purposes. The name denotes the appearance of the material, which somewhat resembles that of a hairy caterpillar. It is usually made of silk, is sometimes a combination of silk and wool, and has been made of wool only."

Colado. A type of embroidery and lace hybrid of Philippine origin, traditionally worked on a fine-weave gauze-like cloth made from pineapple fiber.

Counted thread embroidery. Any of many types of embroidery worked over counted warp and weft yarns of the ground fabric, including canvas embroidery (needlepoint) and cross-stitch. Fabrics used for this purpose are called *even-weaves.*

Crewelwork. "Embroidery with worsteds [fine wool yarns] . . . upon plain materials," such as linen.

Crochet. A method of forming a continuous-weft fabric with a hook and thread or yarn, believed to have been established as a needlework technique in Europe and Britain in the early nineteenth century.

Cross-stitch. Also known as *marking stitch,* a method of crossing one stitch over another on an even-weave fabric to produce a geometrically even design. "Its beauty consists of its points being enclosed in a perfect square."

Cutaways. What remains of a fabric after the parts of a garment or other article to be sewn have been cut from it; the scraps or leftover pieces.

Cutwork. A hybrid of lace and embroidery, in which parts of the ground fabric are cut away after securing the edges with embroidery stitches.

Double-running stitch. A method of forming a line of stitches or a seam by starting with a line of running stitch and then going back over the same line in the opposite direction, so that a complete line of stitches is visible on both sides of the fabric.

Drawnwork. A hybrid of lace and embroidery in which some of the yarn ends of the ground fabric are drawn out and the remaining mesh is secured with embroidery stitches.

Ecru. The natural color of linen, a tan or light brown.

Etui. A small case for holding sewing or embroidery needles, usually cylindrical or square in section.

Farriery. The shoeing and general care of horses and other equines, before the twentieth century typically including veterinary care as well.

Felling. "A term used in sewing. Two pieces of material being first run together [i.e., with running stitch], turn the raw edges over, and hem them double, placing them flat down upon the stuff [the ground fabric]. The turn-over edge should be deeper than that underneath, so the hem may be less bulky, and that the needle employed for hemming may pass through two folds only, instead of four." Also called *flat-felling.*

Filet. Any of a number of kinds of needle- and hook-made laces, in which an even mesh fabric is produced.

Fisgig. A whirligig or top that makes a fizzing noise when spun.

Floss. Embroidery yarn that is spun and put up in hanks or skeins, but not hard-twisted into thread, used to produce a soft finish.

Flower. The design motifs of lace, as opposed to the connecting brides.

Fretsaw. A vertical-bladed saw used for decorative woodcarving.

Gauge. The relationship between the needle or hook size, yarn or thread, and the artisan's working tension that determines how many stitches and rows are produced per inch of knitted or crocheted fabric.

Grain. The warp and weft directions of a fabric. Pattern pieces for clothing must be laid out in the correct grain direction for the garment to fit and drape properly.

Gusset. In the knitting and sewing of garments such as shirts, dresses, and gloves, a small, often wedge-shaped area of fabric that allows for freedom of movement.

Haberdasher. "A seller of trifling wares, such as Tapes, Buttons, Needles, Ribbons, Hooks and Eyes &c., to which articles—all employed in Needlework—the term Haberdashery applies in English."

Hardanger. A type of embroidery combining the elements of cutwork and drawnwork on even-weave fabrics, thought to be Norwegian in origin.

Hedebo. Like hardanger, another hybrid of lace and embroidery techniques, involving cutwork and satin stitches, of Scandinavian origin.

Hemming. "Term used in plain sewing, and the stitch and method of its application . . . to produce a firm, neat border to any article of clothing, upholstery, or of household use, instead of leaving a raw edge, which would ravel out. To make a Hemming, turn in the raw edges of the stuff with a double fold-over, insert the needle, and secure the thread under the fold, and, directing the needle in a slanting position leftwards, take up a couple or three strands of the stuff of single portion, below the fold, bringing the needle through the edge of the fold likewise."

Holland. "A kind of linen . . . only half or altogether unbleached . . . The half-bleached kinds are sized and glazed."

Hot-iron transfer. A type of surface-embroidery pattern introduced at the end of the nineteenth century in which the design is printed in reverse on a tissue-like paper using colored wax, which transfers the design to the fabric when the back of the tissue is pressed with a hot iron.

Housewife (or hussif). A small, pocket-size sewing and mending kit, typically containing thread, needles, patches, and spare buttons, and more rarely a thimble, darning egg, and/or small scissors.

Huckaback (or huck). "A coarse kind of linen [or cotton] cloth, manufactured in small knots at close and regular intervals, making a rough face. It is employed for towels, and is very durable." Popular for counted embroidery in the United States in the first half of the twentieth century.

Imping pins. Shafts sewn to a hawk's or a falcon's broken tail or pinion feathers, to mend the break.

Joinery. The aspect of the making of wooden furniture or other articles involving the joining of two or more wooden parts.

Loft. The fluffiness or puffiness of a fabric; its capacity for trapping dead air.

Mercer. A textile merchant.

Mitt. A piece of handgear made like gloves, but either lacking fingers entirely, or with truncated fingers ending at the first finger joint.

Mountmellick embroidery. A type of raised floral whitework embroidery introduced into Ireland in the early nineteenth century.

Nalbinding. A predecessor of knitting using a hook instead of needles.

Nef. A ship model made of precious metal, typically used as a table decoration.

Palampore. A hand-painted or hand-printed cotton textile of Indian origin, typically decorated with colorful motifs of flowers, trees, and birds.

Picot. "Little Loops or Bobs that ornament Needle-made [and tatted] Laces of all kinds, and that are often introduced into Embroidery."

Piecing. The joining of two pieces of fabric by sewing. In pieced quiltmaking, the assembly of the component blocks prior to quilting.

Point protector. A cap or knob used to hold stitches on a knitting needle.

Pouncing. A method of transferring a design to a fabric using a colored powder called *pounce:* "Rub the Pounce over a piece of paper on which the pattern has been drawn, secure it firmly on the cloth, silk, or velvet, to be embroidered, and prick the pattern through to the material beneath it, so as to deposit the Pounce upon it. Paint the outline with drawing liquid."

Punchwork. A lace and embroidery hybrid in which the fabric is decoratively pierced with a punch, and the resulting holes are then secured with buttonhole and/or blanket stitches to prevent raveling.

Running stitch. "The passing of a needle and thread in and out of the material to be sewn, at regular intervals." A seam sewn with running stitch has the appearance of a dotted line. The interstices can be filled by crossing back across the seam with another line of stitches, to produce double running stitch.

Skein. Yarn or thread wound loosely around a reel or other measure, then twisted together and/or knotted.

Slub. Thick areas in a spun yarn or thread, usually the result of uneven tension in spinning or twisting.

Soutache. "Very narrow silk braids . . . having an openwork centre," used in embroidery.

Stamping. The transfer of a design to fabric by means of woodblock or other printing methods.

Stepping. The creation of the visual effect of a curve, in either counted embroidery or weaving, by making successive "steps" on the fabric grid.

Stillroom. A storage location for dried herbs, preserves, distilled beverages such as cordials, and other botanical *materia medica.*

Strap-hung. A horse-drawn vehicle of the suspension type, predating leaf springs, in which the body of the vehicle is suspended from leather straps called thoroughbraces, attached to a central pole assembly underneath, called the perch.

Surface embroidery. Any of a number of types of embroidery in which the ground fabric is treated simply as a surface, i.e., any embroidery which is not counted work on even-weave fabric.

Tambour embroidery. A method of producing chain stitch embroidery by hooking a series of yarn or thread loops through a stretched fabric.

Tatting. The making of sliding knots with a shuttle and hook, joined to create a delicate lace that is dominated by circular or oval motifs.

Teneriffe. One of the many hybrids of lacemaking and embroidery, believed to have originated in the Canary Islands. Holes are made in the ground fabric, secured with a buttonhole stitch; and wheel-shaped needle lace motifs, resembling the yarn structure known as "Ojo de Dios," are constructed by needle-weaving within the holes.

Thread. The product of plied yarns, twisted tightly in the opposite direction from that of the original spinning of the component yarns.

Tow. The short fibers of the flax plant, valued for their softness and light color, formerly used for the making of hand sewing thread.

Turnery. The making of wooden objects by turning them in a lathe.

Viniagrette. A small container for vinegar used to clean the hands for lacemaking; also a container used for smelling salts.

Warp. The yarns that run longitudinally in a loom, ordinarily attached to the warp beam at the back of the loom, passed through the heddles that lift the yarns and the reed that compacts the warp yarns, and attached to the cloth beam at the front of the loom as the first step in weaving.

Weft. In weaving, the yarn that is perpendicular to the warp and typically carried across the warp in a shuttle, "that is, from selvedge to selvedge, in a web."

Whipstitch. An overcast sewing stitch used to join two pieces of fabric or to secure a hem.

Whitework. Any embroidery done in thread or yarn of the same color as the ground fabric.

Worsted. Wool made of combed long fibers.

Yarn. The product of spinning, either plied or not, but not hard-twisted as in the case of thread.